MW00907692

THE
SOLAR
BOAT
BOOK

BY

PAT RAND ROSE

Ten Speed Press

TEN SPEED PRESS
P.O. Box 7123
Berkeley, California 94707

You may order single copies prepaid direct from the publisher for $8.95 (paper) or $14.95 (cloth), plus $1.00 for postage and handling. (California residents add 6% state sales tax; Bay Area residents add 6½%.)

Library of Congress Catalog Number: 80-69217
ISBN 0-89815-089-2 cloth
ISBN 0-89815-086-8 paper

Cover Design by Brenton Beck
Illustrations by Nancy Austin
Type set by Joanne Shwed. Beverly Anderson Graphic Design.

Printed in the United States of America

10 9 8 7 6 5 4 3 2 1

The Invitation

Come sail with me within these pages. Feel the exhilaration of blue water passing under the bow as *Tevake* slips silently through the waves far out at sea. Feel reborn with me, fresh and new, ready to share the secrets of the Universe. Sail with me in gale force winds and high waves that test our stamina and endurance and make us exult in the power of the sea.

Float in a dead calm. Meet new people while walking new beaches with me. Anchor in new harbors. Beat to windward on a brisk day or run into old friends when returning to old familiar ports.

Most important, come sail with me and experience the freedom and pride of being as energy independent as you choose. Cut that electrical cord that ties you to the land and flourish in an environment of your own choosing.

Let the energy "crisis" pass you by.

Special thanks to:

American Society of Heating, Refrigeration and Air Conditioning Engineers, Inc. for permission to adapt and reprint tables from the 1977 Fundamentals Volume, ASHRAE Handbook and Product Directory.

E.W. Bottum, Solar Research Division of Refrigeration Research, Inc., for his assistance in researching solar refrigeration.

The Cousteau Society, Inc., for permission to reprint excerpts from the keynote address by Jacques-Yves Cousteau at the Seattle, Washington Involvement Day, October 1977.

The Mother Earth News for its inspiration and continued interest in solar research and development.

My brother, Dr. James L. Rand, of SW Research Institute, San Antonio, Texas, for his assistance with my research on photovoltaics.

Ed Walters, my friend.

Contents

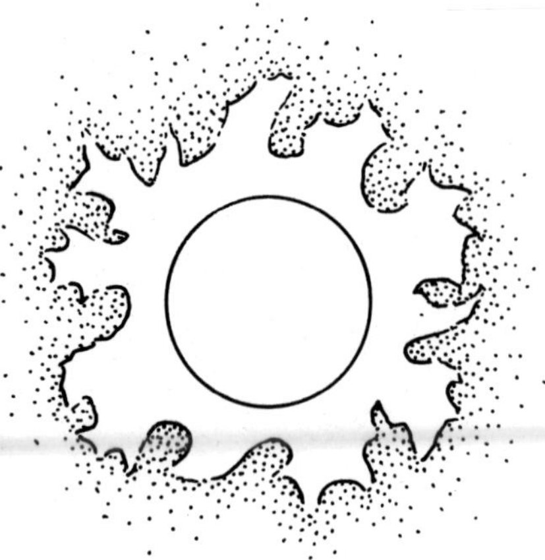

1
The Basics — Those Forgettable Facts

The amount of solar energy that falls on Lake Erie alone, is enough to supply all the energy needs of this country.[1] But just how much is that? And how do we capture it? And what do we do with it after we have it? That is exactly what this chapter is about.

Before you design a solar energy system for your boat, you will need to re-acquaint yourself with certain fundamentals. You will need to know something about the sun, the climate, the nature of heat, the properties of some building materials and something about solar collectors. These basics will be the tools you need to help you decide which systems are best for your boat. You probably were exposed to some of this information by hard-working teachers during early school days, and some of it may still be with you. But for the benefit of those, like myself, who considered school an obstacle to their social development, a brief review follows.

The sun, that rotating mass of burning gases radiating energy at unbelievable rates, is the center of our solar system. It is hard at work, as it has been for five billion years, turning millions of tons of matter into energy every second. Although the sun seems to move across our heavens, the earth actually is moving counterclockwise around the sun, following an oval path. At the same time as the earth moves around the sun, it is turning on its own axis, wobbling along, counterclockwise again, and with a decided list. This list is responsible for many things. It causes the north pole to be tilted away from the sun in the winter, so that the sun appears lower on the horizon. And even though the earth is closest to the sun at this time, the sun's rays must strike the earth a more glancing blow. This results in a smaller amount of solar radiation striking a horizontal surface, consequently colder weather. The same listing rotation that causes it to be cold in the winter causes the heat of summer and is responsible for the change of seasons.

In the northern hemisphere, the sun is highest in the sky June 22 and lowest on December 22, with midpoints at the equinoxes, March 21 and September 23.

Most meteorological stations report solar radiation in terms of total langleys received on a horizontal surface at groundlevel. One langley per minute = one calorie of radiation energy per square centimeter per minute, or 221 BTUs per square foot per hour.

Part of this groundlevel radiation may come directly from the sun, but often a large percentage will come from scattered or reflected light, especially near water. When the sky is clear, the solar radiation striking a horizontal surface is greatest at the equator at noon. At all latitudes, the sun moves from east to west through an arc of 15 degrees each hour. In the early morning and late afternoon, the sun's rays must travel farther

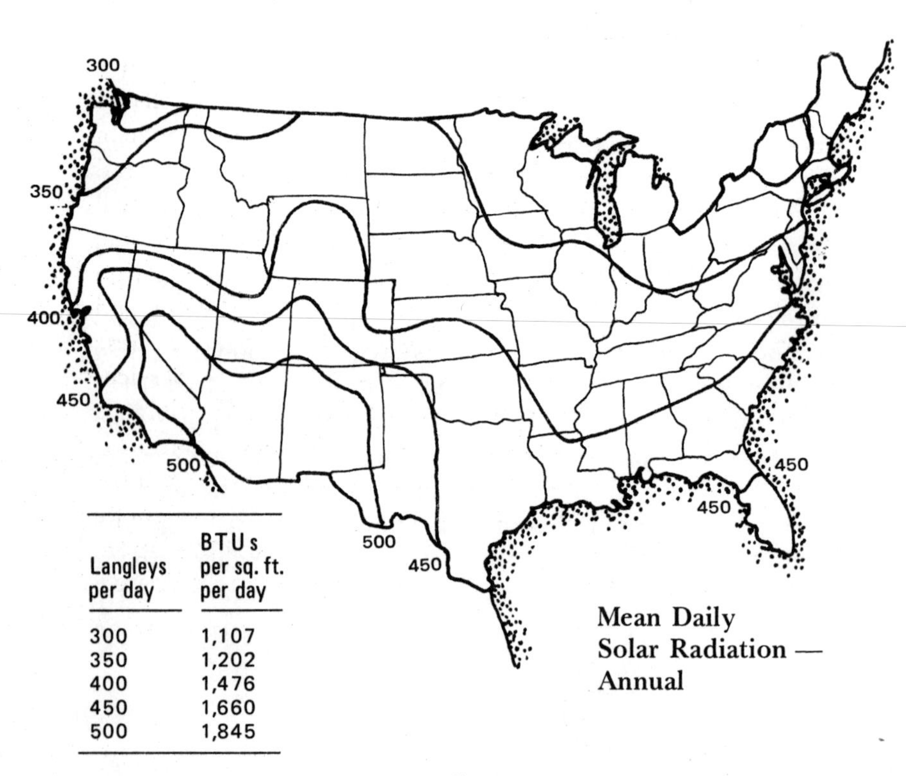

Langleys per day	BTUs per sq. ft. per day
300	1,107
350	1,202
400	1,476
450	1,660
500	1,845

Mean Daily Solar Radiation — Annual

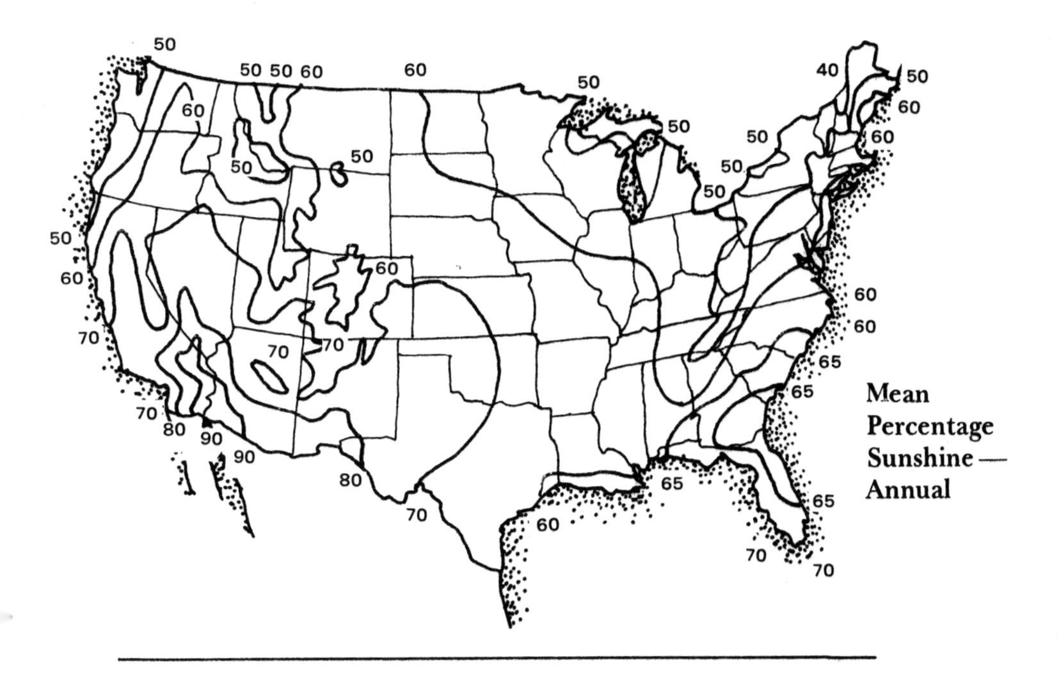

Mean Percentage Sunshine — Annual

and more obliquely; therefore, radiation intensity depends on where you are, what time of day it is, the season, clouds and the amount and types of air pollution.

The sun maps shown here are reprinted to give you a rough idea of the huge amount of solar radiation that falls on *all* parts of the country. The energy falling on your boat in mid-winter, regardless of its geographical location, is several times the amount of energy needed inside the boat. We will learn how to get that energy from the outside to the inside, and how to keep it there. If you would like more information on the solar radiation in your area, consult the tables and maps in the latest Climatic Atlas of the United States. These data are assembled by the Environmental Science Services Administration and can be found in your local public library.

Now that we are aware of how much solar radiation is being received, we need to learn about another type of radiation. Every warm body gives off an aura, or electromagnetic radiation. Sunlight is one form of electromagnetic radiation, heat or thermal radiation is another. They differ only in wave length.

Table 1

RESISTANCE OF COMMON INSULATION MATERIALS

MATERIAL	THICK-NESS	R-VALUE*
Air space – Horizontal	3/4 inch	0.76 summer, 0.87 winter
Air space – Horizontal	4 inch	0.80 summer, 0.94 winter
Air space – Vertical	3/4 inch	0.84 summer, 1.01 winter
Air space – Vertical	4 inch	0.91 summer, 1.01 winter
Carpet with fiber pad	standard	2.08 R-value
Carpet with rubber pad	standard	1.23
Concrete – gypsum fiber	8 inch	4.80
Concrete – sand and gravel	8 inches	0.88
Corkboard – 90° F.	1 inch	3.57
Cork tiles	1/8 inch	0.28
Fiberglass board	1 inch	4.35
Glass – cellular – 90° F.	1 inch	2.44
Glass – fiber – 90° F.	1 inch	3.85
Glass – flat	1/8 inch	0.89
Glass – inculating		1.54
Cellulose fiber, loose fill insul.	1 inch	3.70
Macerated paper insulation	1 inch	3.57
Mineral wool insulation	1 inch	4.00
Sawdust insulation	1 inch	2.22
Mineral wool batt	3-1/2 inch	10.90
Mineral wool batt	6 inch	18.80
Plywood	1/4 inch	0.31
Plywood	3/8 inch	0.47
Plywood	1/2 inch	0.62
Plywood	5/8 inch	0.78
Plywood	3/4 inch	0.94
Polystyrene, expanded, 75° F	1 inch	3.85
Polystyrene, molded, 75° F	1 inch	3.57
Polyurethane, expanded, 75° F	1 inch	5.88
Tiles – asphalt	standard	0.05
Tiles – linoleum	standard	0.05
Tiles – rubber	standard	0.05
Tiles – vinyl	standard	0.05
Wood – hardwoods	1 inch	0.91
Wood – softwoods	1 inch	1.25

*The higher the R-value of a material, the more resistance that material has to the flow of heat; therefore, the better insulator it is.

Heat is nervous by nature. It is constantly in motion, moving from warm places toward colder ones in its constant effort to equalize the temperature. The greater the difference between temperatures, the faster heat moves. The amount of heat that flows, however, depends on the amount of resistance there is to that flow. Insulation offers resistance to heat flow. The Insulation Table (Table 1) was excerpted from ASHRAE, Handbook of Fundamentals, 1977, and shows only the insulating materials you might consider using on your boat. The R-value shown on the table represents the amount of resistance of each particular material to heat flow. The higher the R-value, the better insulator the material is.

Heat flows by three basic methods – conduction, convection and radiation. The flow of heat by *conduction* is the transfer of heat from its host to a material that is in direct contact with the host. When the handle of a skillet sitting on a stove heats up, this is an example of conductive heat flow. Heat has been conducted from the stove, through the skillet, to the handle.

Heat flows by *convection* when it is transferred to liquids or gases, which then expand, become more buoyant, and rise, warming the surrounding area. A person on deck on a cold, windy day gives up body heat by convection, and the convective heat loss increases as the wind velocity increases and as the difference in temperature between his body and the air increases. In this instance, wearing protective clothing that gives resistance to heat loss can be very important.

Heat flow by *radiation* is most important when transferring heat from one place to another. When radiation strikes any surface, it is reflected, absorbed, and transmitted according to the characteristics of the receiver and the wave length

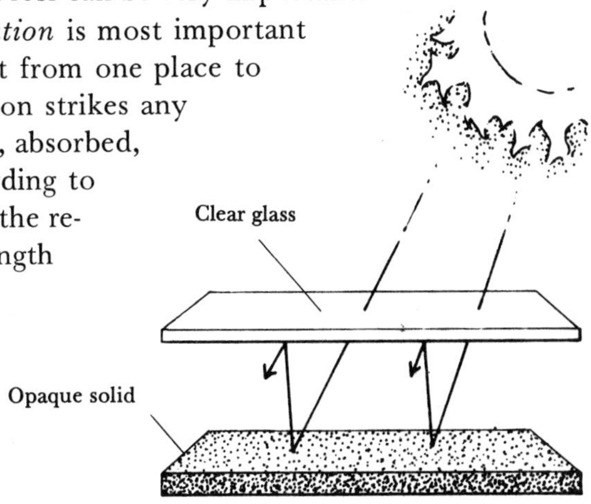

Clear glass

Opaque solid

Green House Effect

of the radiation. Glass, for instance, allows the sunlight to pass through it (transmits most of the solar radiation that hits it), while absorbing most of the heat (thermal radiation). This is called the "greenhouse effect" and will be very important in developing successful solar heating for your boat. An opaque solid, on the other hand, allows no sunlight to pass through it; therefore, it transmits no energy, but reflects away the energy it has absorbed.

Knowing something about the propensity of different materials to absorb and emit heat can help you select the best materials for the job when modifying your boat or constructing solar collectors. Numerical ratings are assigned to gauge the ratio of radiation absorbed to the total solar energy available. Once this solar energy is absorbed, the radiant energy is transformed into heat energy. The material becomes warmer and emits more radiation of its own. The emittance is a

Table 2

HEAT ABSORBTANCE AND EMITTANCE

MATERIAL	ABSORBTANCE	EMITTANCE
Aluminum foil	.15	.05
Aluminum paint – oil base	.45	.90
White paint on aluminum	.20	.91
Black paint on aluminum	.94 – .98	.88
Cupric oxide on aluminum	.85	.11
White enamel on iron	.25 – .45	.90
Galvanized sheet iron, oxidized	.80	.28
Galvanized sheet iron, bright	.65	.13
Nickel black on galvanized iron	.89	.12
Dull brass, copper, or lead	.20 – .40	.40 – .65
Grey paint	.75	.95
Green oil base paint	.50	.90
Red oil base paint	.74	.90
Black gloss paint	.90	.90
Lamp black	.98	.95
Sand – dry	.82	.90
Sand – wet	.91	.95
Water	.94	.95 – .96

numerical rating of the material's propensity to radiate away its energy. The Absorbtance and Emittance Table (Table 2) is excerpted from ASHRAE, Handbook of Fundamentals, 1977.

Table 3 shows the capacity of certain materials to absorb and *hold* heat. The "Density" column shows the weight of a

Table 3

HEAT STORAGE CAPACITY

MATERIAL	DENSITY *	HEAT CAPACITY **
Air	0.075	0.018
Aluminum – Alloy 1100	171	36.6
Asbestos fiber	150	37.5
Asbestos insulation	36	7.2
Brass – Red – 85% copper	548	49.3
Brass – Yellow – 65% copper	519	46.7
Bronze	530	55.1
Cellulose	3.4	1.1
Cement – Portland Clinker	120	19.2
Concrete – stone	144	31.7
Copper – electrolytic	556	51.2
Cork – granulated	5.4	2.6
Fireclay brick	112	22.2
Glass,– crown – soda-lime	154	27.7
Glass – flint – lead	267	31.2
Glass – Pyrex	139	27.8
Glass wool	3.25	0.5
Iron – cast	450	54.0
Lead	707	21.8
Salt – rock	136	29.8
Steel – mild	489	58.7
Water – fresh	62.4	62.4
Water – salt	72	54.0
Wood – hardwoods	47	26.8
Wood – softwoods	27	17.6

* Weight in pounds of one cubic foot of material.

** Amount of heat in BTUs absorbed by one cubic foot of the material during a one degree rise in temperature.

cubic foot of each material, and the "Heat Capacity" column shows the amount of heat (in BTUs) absorbed by one cubic foot of each material during a 1 degree F rise in temperature. This table will help you select the right materials for storing the heat you collect. Study it carefully. You may be surprised at the enormous heat storage capacity of some very common mediums – for example, water.

A solar collector is the link between the sun and the solar boat. There are two basic kinds of *thermal* solar collectors: the flat plate collector and the concentrating collector.

The *flat plate collector* is made of sheet metal (usually iron, copper or aluminum) to give good heat conduction. The surfaces are blackened with a dull paint containing carbon black, or a chemically produced black coating. The plate absorbs radiation. Then, to prevent the collector from wasting heat by giving it off to its surroundings, one or more pieces of glass or clear plastic are placed over the top of the collector, trapping the heat inside, where it can be transferred to the working fluid (usually air or water) flowing on one side of the collector.

A flat plate collector is very simple to build. The blackened absorbing plate is placed inside a frame, which should be made of thin wood, plastic or very thin metal to minimize heat loss by thermal conduction. For heating water, the water pipes are attached (ensuring good thermal contact) to the back. For heating air, an air duct is provided behind the blackened sheet. The box is then heavily insulated with glass wool, plastic foam, or some other material that has a high R-value, and a sheet of glass or transparent plastic is placed over the front of the collector to prevent heat loss from convection and radiation. A collector can be made in any size or shape to best suit your needs and your available space.

Maximum energy would be obtained if the flat plate collector were tilted so that it was always at right angles to the sun, which is moving 15 degrees each hour. However, this is usually impractical, especially aboard a boat. For land applications, the collector is normally mounted facing the equator, tilted at an angle equal to the latitude plus 15 degrees. This

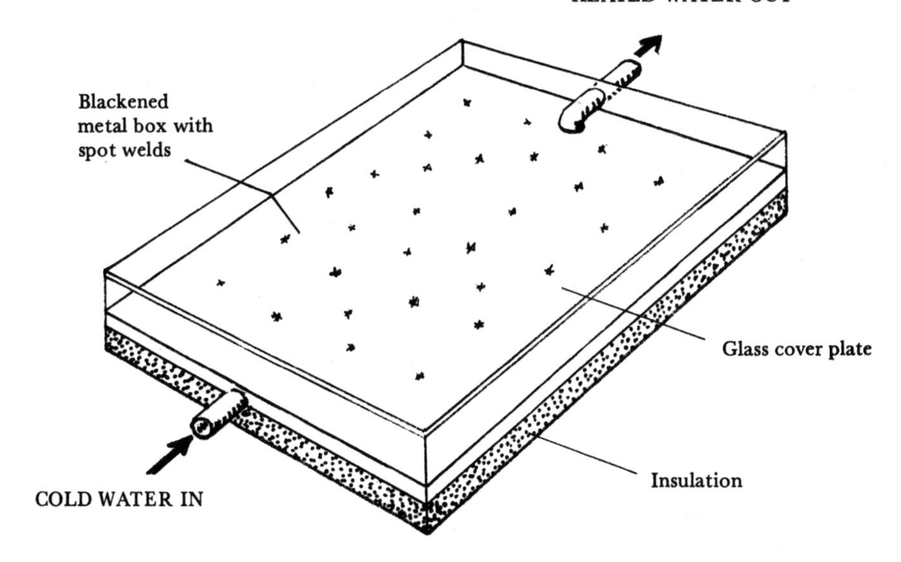

Flat Plate Collector For Heating Water

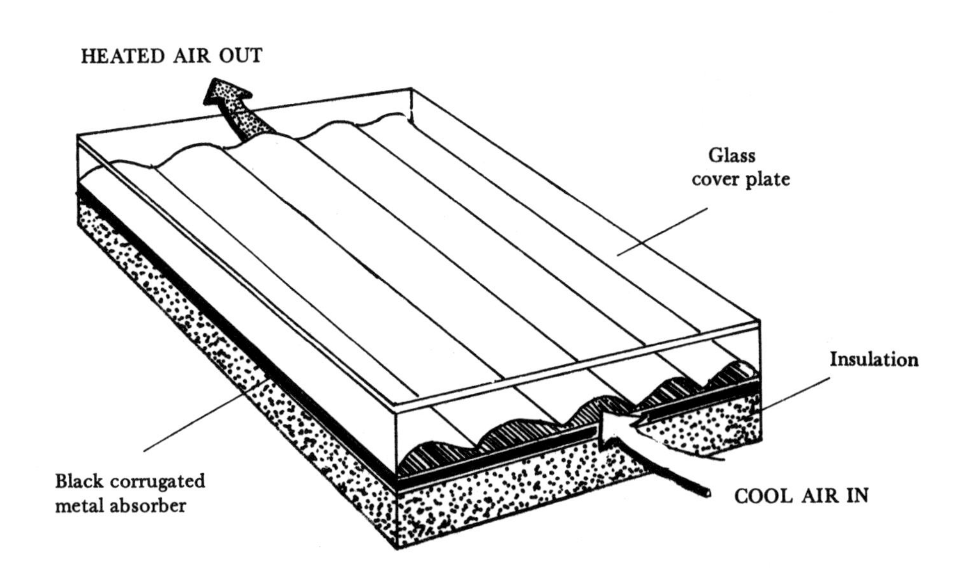

Flat Plate Collector For Heating Air

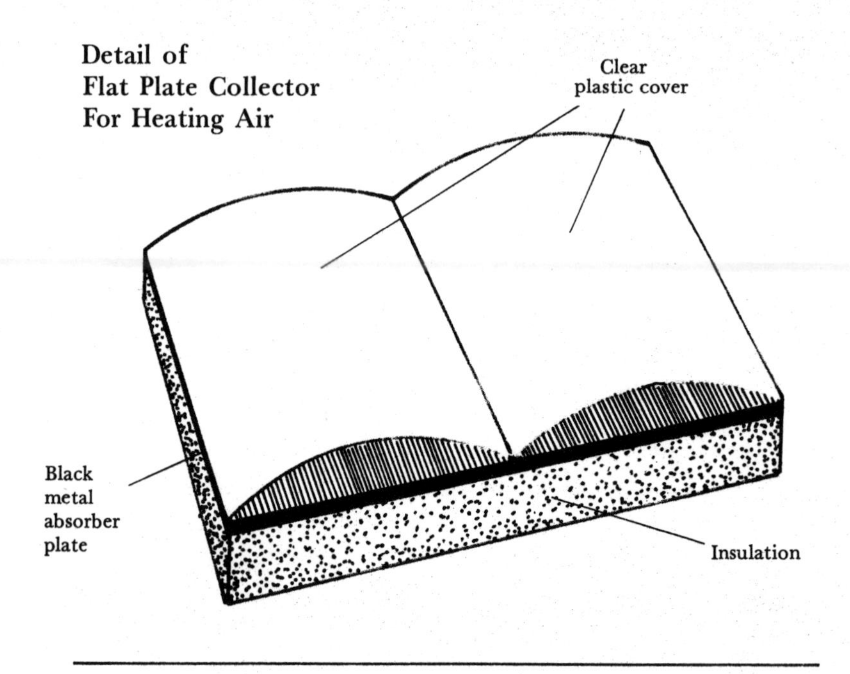

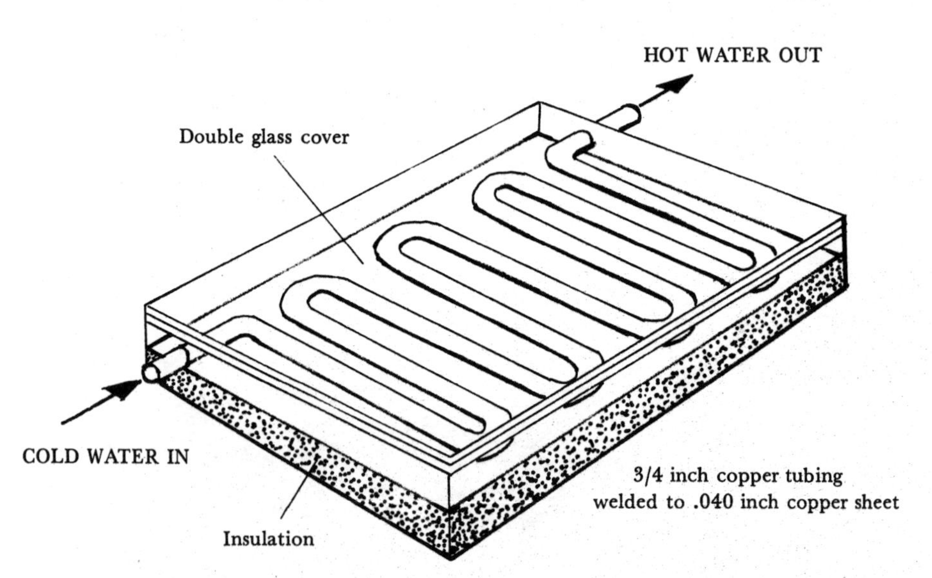

Detail of Flat Plate Collector For Heating Water

will result in the highest temperatures possible for a fixed flat plate collector. Aboard your boat, however, you have to mount the collector wherever it is most convenient. The amount of solar radiation you collect will then vary slightly according to the time of day, the season and the orientation of your boat. Flat plate collectors are capable of boiling water, but they seldom achieve temperatures higher than 200 degrees F.[2]

A *concentrating collector* is used where extremely high temperatures are desired. Its main disadvantage is that it *must* be rotated to track the sun in order to attain the high temperatures for which it is designed. This makes it suitable for shipboard use only in short-term applications (such as making coffee or heating a pot of soup) while at anchor or on a long single tack. Full-time use of this type collector would require that your boat have a more permanent orientation to the sun; it would be completely satisfactory only when you are tied up at the dock.

A concentrating collector uses one or more reflecting surfaces to concentrate sunlight onto a small absorber. The concentrating collector can be made in several shapes: parabolic or spherical, circular or cylindrical. A parabolically shaped concentrating collector is curved to focus sunlight onto a very small area. The sharper the focus of the collector, the higher the temperature that is generated. This sharp focus, concentrating the intensity of the sun's rays, is capable of temperatures of 3,000 degrees C.

A spherically shaped collector can be used instead of a parabolically shaped collector if a sharp focus and the resulting extremely high temperature aren't needed. A spherical collector gives a fairly good focus and still yields a fairly high temperature.

Circular collectors yield the highest temperatures, but cylindrical collectors, while yielding slightly lower temperatures, can be mounted in a manner that requires a minimum of tracking of the sun.

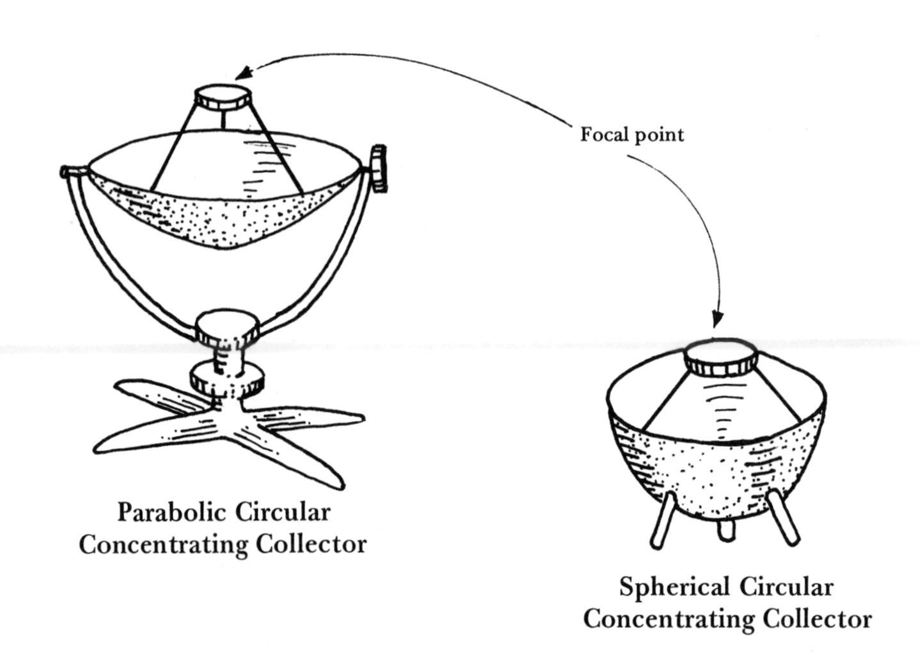

Parabolic Circular Concentrating Collector

Spherical Circular Concentrating Collector

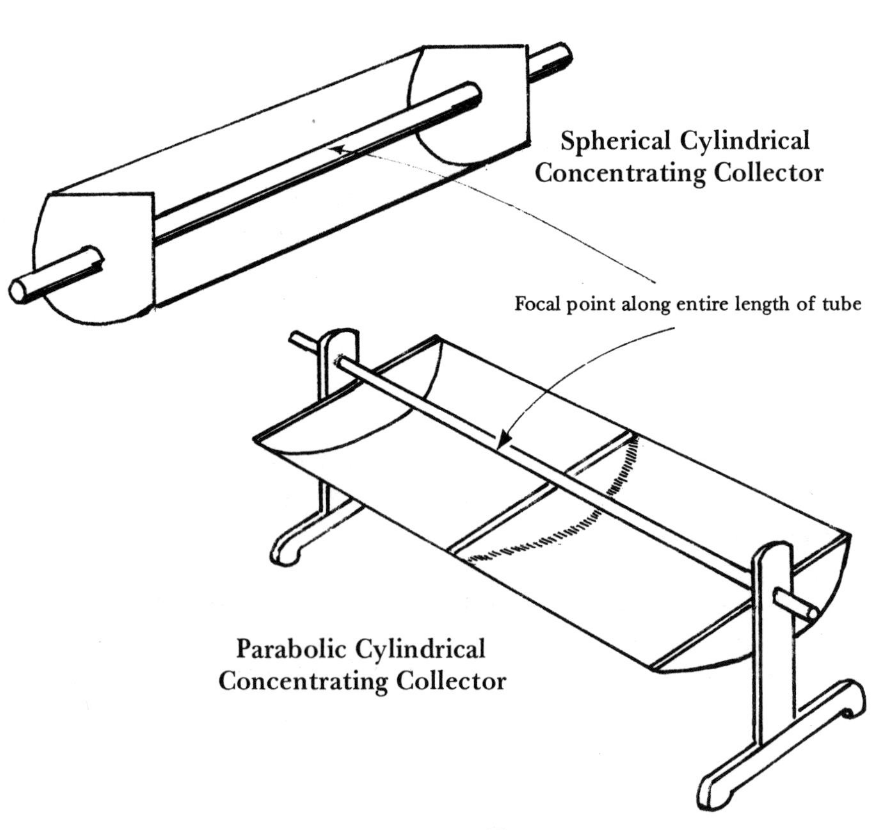

Spherical Cylindrical Concentrating Collector

Parabolic Cylindrical Concentrating Collector

The long axis of the cylindrical collector can be aligned east to west and adjusted daily or weekly so that the center target strip is in focus. The receiver will be in focus most of the day, except for early morning and late afternoon changes. Mirrored or aluminized reflectors can be added, which will gather and reflect additional sunlight onto the absorbers.

Another type of concentrating collector can be made by arranging reflective surfaces in a manner that will gather sunlight and bounce it onto an absorbing surface as shown.

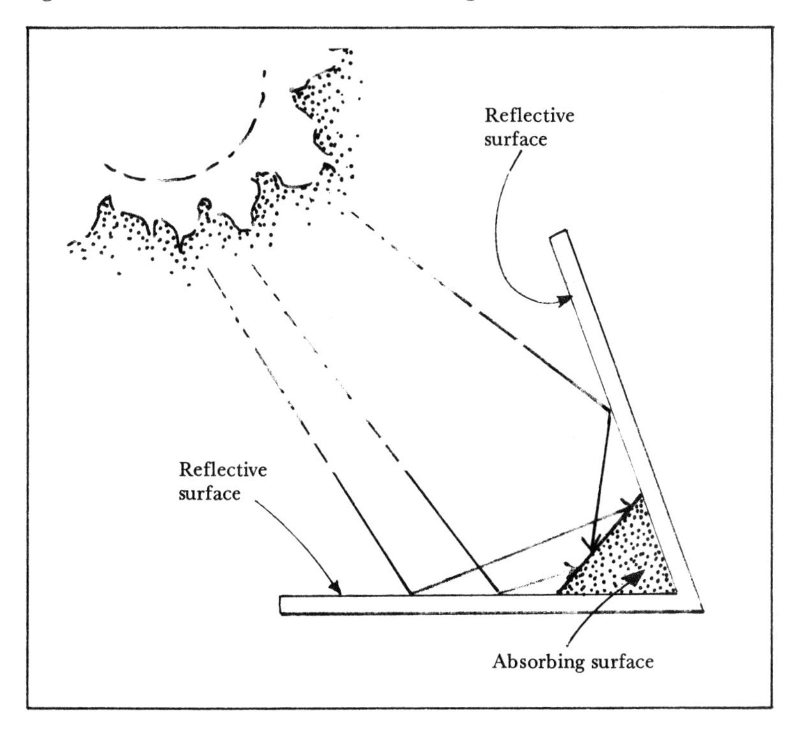

Another device for focusing solar energy onto a target area is the fresnel lens. Inexpensive fresnel lenses are now available. They are made of a lightweight plastic called acetate butyrate. A series of concentric lines are etched into the plastic a few thousandths of an inch apart. Each line acts as part of a lens, and taken together, they gather light and focus it into a single spot, much like a giant magnifying glass.

A fresnel lens is so powerful that any combustible material, such as paper or wood, will immediately burst into flame when placed at the focal point of the lens. Reflection and glare at this point is also dangerous to the eyes. Dark glasses should be worn at all times when focusing the lens, in order to prevent retinal burns or other eye injuries. Extreme caution must be exercised when experimenting with a fresnel lens.[3]

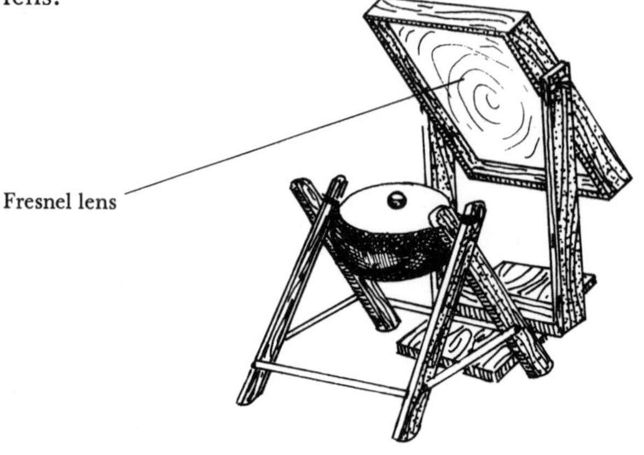

A completely different kind of link between the sun and the solar boat is the photovoltaic collector. Instead of converting the sun's rays to heat, the photovoltaic collector converts sunlight to electricity. A solar (photovoltaic) cell has a positive layer containing movable positive "holes," and an N-layer containing movable negative electrons, with a barrier in between. When light enters the cell, electrons are released and flow to an electrode and through a wire to the other electrode, where they combine with the positive holes. The barrier layer prevents the instant recombination of electrons and positive holes, causing electrons to flow through the wire, generating electricity. These solar cells are wired together in series to produce a photovoltaic collector of the appropriate voltage.[4]

The basics discussed in this chapter have introduced you to the tools available for collecting and using the energy of the sun. More specific information on these tools and how to use them is contained in the chapters to follow.

2
Evaluating Your Electrical Needs

Since much of the energy used aboard the modern boat is electrical, we must study your electrical needs carefully. If you have been boating for even a short time, you probably have had some intimate experience with your electrical system already.

If you have ever had your running lights fade from dim to nonexistent twelve miles offshore in the middle of a shipping lane . . . *if* you have ever tried to start an engine and heard nothing but a dull thud just before being swept onto a sand bar . . . *if* you have ever removed the pressure system pump and taken it for repairs, only to discover it worked perfectly in the shop . . . you know exactly how vital an efficient electrical system can be.

That dead battery could have been the result of a faulty generator, a loose connection, or even old age, but chances are it was the result of trying to use more electricity than the battery could possibly store. We often use or add more and more equipment over a period of time without sufficient thought to the power requirements it adds. The surprises that result from this lack of planning remind us to re-evaluate our power requirements and make the necessary adjustments.

When solarizing your boat, it will be *even more* important to be energy-conscious. If you are to design a really workable solar energy system for your boat and prevent those unpleasant surprises, it will be necessary to evaluate each and every piece of equipment aboard your boat and determine exactly how much energy that equipment uses. Also, a great deal of thought should be given, not only to whether or not each device can justify the amount of electricity required to operate it, but whether or not supplying that electricity will require the expenditure of additional fossil fuel.

Sometimes an alternate may be used. If an alternate is not desirable, we can devise ways to conserve the energy we must use, and a regular maintenance program will keep all electrical devices working at peak efficiency. The emphasis must be on using the best fuel for each job. If we cannot totally eliminate our demands for fossil fuels, we *must reduce* these demands wherever it is possible to do so, *without reducing* our standards. This is what Amory B. Lovins, of Friends of the Earth, describes as meeting our needs with an "elegant frugality of energy supplies in the most appropriate way for each task."[1] In other words, we shouldn't use a cannon to kill a fly when a swatter would do the job quite well.

The main electrical system aboard most boats today is 12 volts DC, with the electricity contained in a bank of storage batteries. These storage batteries are usually recharged by an alternator/generator that is belted to the boat's primary engine. A built-in 115-volt AC system is often included for dockside use. We will evaluate both electrical systems on board.

At the right we have provided a sample chart to help you determine your present power consumption. Using this chart as a guide, list every item on board your boat that consumes electricity, noting whether it is AC or DC and whether it is used dockside or offshore. Complete your chart carefully; you will find it your most useful tool in modifying your energy requirements and you will refer to it again and again in the months to come.

List all the DC-powered electrical devices you have aboard your boat first. AC-powered devices can be listed on a separate chart. These will be reviewed in Chapter 10. It will be helpful if you group these devices under the broad headings of "Safety" (such as radio, engine start, bilge pump, etc.), "Navigation" (such as autopilot, depth sounder, knotmeter, etc.), "Comfort" (such as heating, fans, refrigerator, etc.), and "Luxury" (such as TV, stereo, etc.). This will be useful later when deciding which items to limit use of during crucial times.

ELECTRICAL CONSUMPTION CHART

Appliance	AC	DC	Used Off-Shore	Used Dock-Side	Volts	Watts	Ohms	Amps	Hours Used Daily	Daily Amp Hours Used	Possible Substitute	Remarks

Most electrical appliances have the voltage, amperes, ohms or wattage information printed right on the appliance itself. If not, check the maintenance handbook or operating instructions that came with the item. If necessary, an inexpensive electrician's multimeter can be used to measure volts, ohms, or DC amperes.[2] It is important to be as accurate as possible when compiling this information. Also, since not all manufacturers express the electrical requirements of their appliances in the same terms, it may be necessary to convert your calculations to amperes for the purposes of your Electrical Consumption chart.

$$\text{Voltage} = \text{Amperes} \times \text{Ohms}$$

therefore

$$\text{Amperes} = \frac{\text{Voltage}}{\text{Ohms}}$$

Also:

$$\text{Wattage} = \text{Voltage} \times \text{Amperes}$$

therefore

$$\text{Amperes} = \frac{\text{Wattage}}{\text{Voltage}}$$

One example would be a cabin heater that draws 115 volts, with a resistance of 23 ohms. You can determine the amperes used by following the second equation above:

$$\text{Amperes} = \frac{115 \text{ Volts}}{23 \text{ Ohms}} \quad \text{or 5 Amperes}$$

If the wattage is known, such as in a 15-watt light bulb operating on a 12-volt circuit, the amperes used can be determined by using the last equation above:

$$\text{Amperes} = \frac{15 \text{ Watts}}{12 \text{ Volts}} \quad \text{or 1.25 Amperes}$$

When completing your Electrical Consumption chart, first fill in all the information you have readily available on your electrical devices, then go back over the chart and convert all the electrical information into amperes.

In completing the column marked "Hours Used Daily" you should make the best possible estimate of how many hours you use each device each day. Don't rush to complete the chart. Solarizing your boat is an important project, and the more accurate your original computations, the more satisfied you will be with the end results.

Multiply the "Hours Used Daily" by the "Amperes" for each appliance, and enter the result in the "Daily Amp Hours Used" column. Once the chart is complete, total the "Daily Amp Hours Used" for all DC equipment. This will give you the total drain on your battery bank for a typical day at sea. The next step will be to determine how much amperage your batteries can store.

ELECTRICAL CONSUMPTION CHART

Volts	Watts	Ohms	Amps	Hours Used Daily	Daily Amp Hours Used	Possible Subst
			3	.25	.75	
			.1	1	.1	
			5	12	60	
			7	.5	3.5	
				Total-	64.35 hrs.	

The most popular battery aboard the modern boat is the lead-acid type, identical in configuration to the battery used in most cars, and usually 12 volts. A single cell of the lead-acid storage battery has a capacity of 2 volts no matter how large or small the cell is. The larger cells can generate more amperes, but the voltage remains the same – about 2 volts. If more voltage is needed, the 2-volt cells are connected *in series.* A 6-volt battery consists of three cells connected in series; a 12-volt battery has six cells connnected in series, etc.

The amperage of a cell depends on the amount of surface of lead plate exposed to the electrolyte in the cell. In order to increase that surface, thus increasing the amperage, several plates can be used or the lead plates can be grooved, corrugated or sponged, exposing even more surface to the electrolyte. This latter technique produces a high amperage cell without making the size of the cell prohibitive. However, voltage remains at about 2 volts.

There are several ways to increase available amperage aboard your boat. One way is by adding batteries of identical voltage, connected in *parallel.* The voltage will remain the same, but the ampere hour capacity of the bank of batteries will be the *sum* of the ampere hour rating of all the batteries. That is, if you have a bank of three 12-volt batteries connected in *parallel,* and each battery has an ampere hour rating of 80, your battery bank will theoretically store 240 ampere hours of electricity at 12 volts.[3]

If you have a large cruising boat, where space and weight are not crucial factors, some thought should be given to the use of large, industrial-type batteries. These 2-volt units are available from the industrial division of any battery manufacturer: Gould, Exide, Delco, and many others. These batteries are quite heavy (up to 70 lbs. each), must be wired in series to obtain the appropriate voltage, and cost several hundred dollars for a complete set. But over a long period of time, they will probably be the most efficient and economical way to go. They are sturdily built to withstand the rugged demands of frequent discharge and recharging. They have mammoth ampere hour ratings and a life expectancy of 25 to 30 years.[4]

Golf cart and heavy duty truck batteries have some of the same advantages as industrial batteries. They will give you a more efficient battery bank than ordinary marine or automobile batteries, *if* your boat can accommodate their additional size and weight.

TAKE CARE OF THOSE BATTERIES

For the best possible performance from your storage batteries, whatever type you use, they must be properly installed and maintained.

1. Batteries should be mounted in a dry, well-ventilated area so that hydrogen gas given off during charging will dissipate quickly. They also should be secured against shifting or pounding in rough seas and covered to prevent accidental shorting of the terminals.

2. The top of each battery should be kept as clean and dry as possible, since moisture can cause the battery to discharge. Baking soda is a good cleaning agent for the top of the battery but must never be allowed to enter the cells. After cleaning, the battery should be rinsed and thoroughly dried.

3. The voltage regulator should be checked annually. A charging circuit of 14 volts for power boats and 15 volts for sailboats with auxiliaries should be adequate. Higher voltage could lead to overcharging and lower voltage could allow the batteries to run down.

4. Cable clamps and battery posts should be cleaned with a stainless steel brush or coarse sandpaper before a connection is made.

5. Battery cables should be replaced at the first sign of corrosion or fraying to prevent excessive loss of voltage.

6. Cells should be topped off with the correct level of water. Tap water is fine unless it contains high levels of iron or chlorine.

Monitor the state of charge of your batteries occasionally. This can be done with an hydrometer or with a permanently installed battery condition meter (an expanded scale voltmeter reading between 10 and 15 volts).

Even though your batteries are perfectly installed and maintained, there are additional factors that affect a battery's actual capacity, such as discharge rate and specific gravity. Therefore, the actual ampere hour capacity of your battery may be somewhat different from the theoretical rating.

The higher the amperes of the discharge rate, the less total ampere hours a battery will deliver. This means that any battery of a given capacity will do more useful work operating a light bulb than in cranking a balky engine.

The specific gravity of a battery's electrolyte varies with different brands, but is generally given as 1.260 for a fully charged cell, 1.197 for a half charge, and 1.135 for a discharged cell. In other words, the specific gravity decreases as the battery discharges and increases as it is charged. But if the ambient temperature in the battery compartment is above or below 77 degrees F, or if the electrolyte level is above or below normal, corrections must be applied as follows:

- Add .001 for each 3 degrees F above 77 degrees F.
  Subtract .001 for each 3 degrees F below 77 degrees F.

- Add .015 for each 1/2 inch of water above normal.
  Subtract .015 for each 1/3 inch of water below normal.

An example would be a battery compartment with a temperature of 107 degrees F. The water in the battery is 1/2 inch above normal and the hydrometer reads 1.235. The correction would be a plus .025, making the corrected hydrometer reading 1.260, indicating the battery is fully charged.[5]

The manufacturer's ampere hour rating of your battery may be printed somewhere on the battery itself, but chances are you will find it necessary to check with the dealer or the manufacturer himself. After you have computed approximately how many ampere hours your battery will store, compare this figure to the total daily DC ampere hours used (taken from your Electrical Consumption chart). If the daily total used equals or exceeds the estimated capacity of your battery bank, you know immediately that you must re-evaluate every piece of electrical equipment on board.

Even if your total daily load is less than capacity, charging batteries with an engine-driven alternator/generator is a noisy, smelly, time-consuming operation that uses fossil fuel which might be conserved for a more important task. What better place could there be to begin solarizing your boat than with a solar generator?

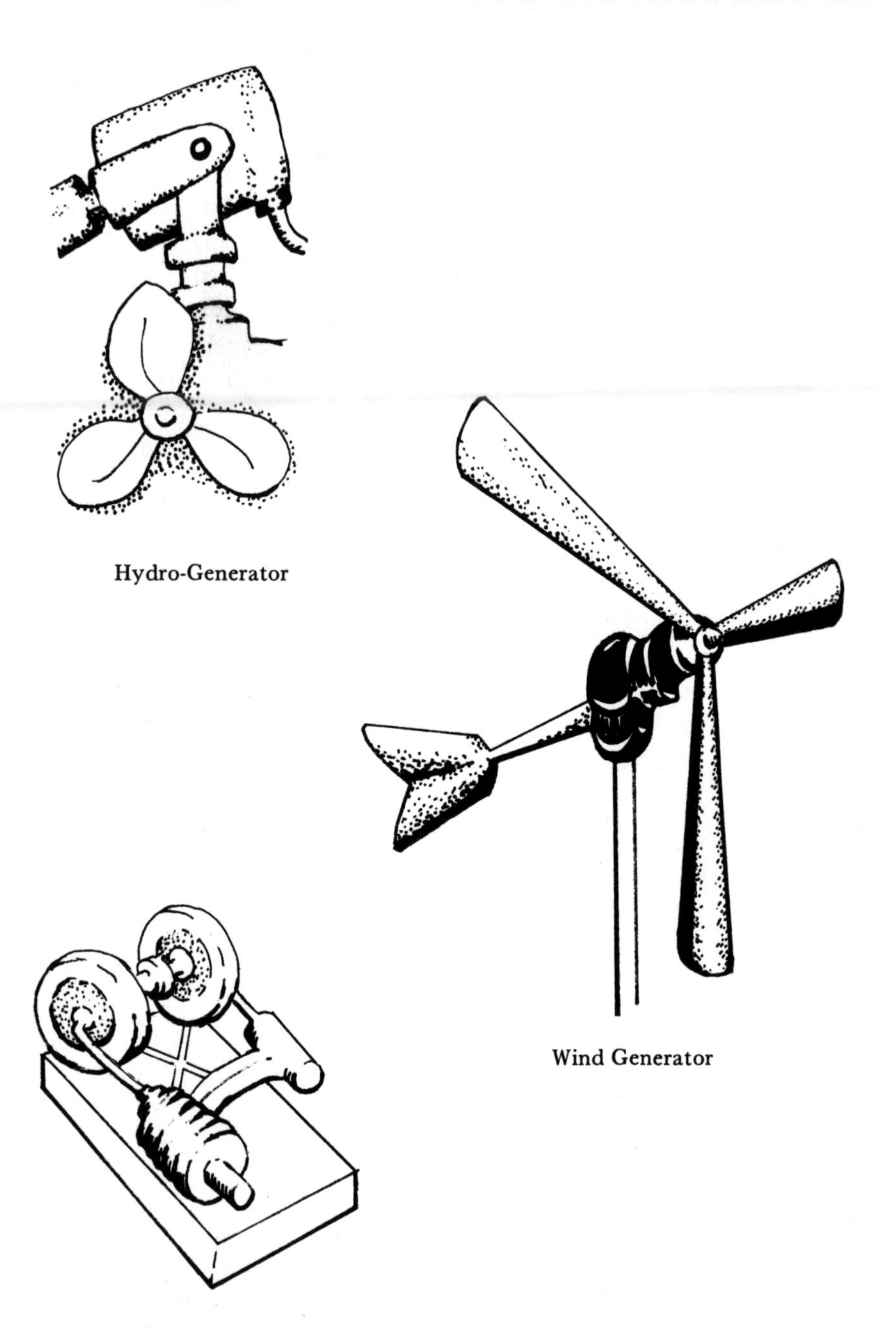

Hydro-Generator

Wind Generator

Stirling Cycle Engine

Free Fuel Generators

3
Free Fuel Generators Sun–Wind–Water

We are all familiar with the fossil-fuel-powered alternator/generator which either runs off the boat's main engine or is powered by its own independent engine. How would you like to charge your boat's battery bank by other methods, using a fuel that is free? The sun, wind, and water – or all three – can be used to generate electricity. The conventional generator can be retained for use only as a backup for times when your electrical demands are extraordinarily high, or when the engine already is running.

The photovoltaic generator is unique, in that it involves no generator at all as we know it. It consists of a series of photovoltaic cells that react to light by producing electricity directly. The development of the solar (photovoltaic) cell made it possible for us to realize the dream of space travel. In the past, the high cost of these cells seemed to be their only disadvantage. However, the Department of Energy's search for radically less costly solar cells has resulted in new manufacturing techniques that are causing prices to plummet.

The Solar Photovoltaic Energy, Research, Development and Demonstration Act of 1978 authorized a ten-year, $1.5 billion program to reduce the cost of photovoltaics to the point that they would be cost-competitive with utility-generated electricity. The cost goals are now $2.80 per peak watt by the end of 1982, 70 cents per peak watt by 1986, and 15 cents per peak watt by 1990. Also, solar panels must meet requirements of 10% efficiency and a life of 20 years. Since flat panel *silicon* cell arrays held the most promise, research has been concentrated on improving the techniques for the manufacture of the silicon cell.

A solar cell begins with a pile of sand. Silicon, the second most abundant element on earth, is contained in that pile of sand. Extracting the pure, high-grade silicon from that sand is the first problem. The cost of the silicon is from 25% to 50% of the price of the solar array, and has been $45-100 per kilo. To meet the DOE's cost goals, a price of $14 a kilo is necessary, and several suppliers are implementing new processes for the production of high-grade silicon that will soon meet this goal.

There are now four methods for producing the cells from high-purity silicon, three of which have only recently been put into commercial operation.

The traditional method for making solar cells is still being used by many manufacturers who have improved automation to bring the price of their products down. It consists of melting the purified silicon at 1450 degrees C in a quartz crucible. A "seed" crystal is dipped into the melt and slowly pulled up to form a sausage-shaped single crystal. It is then cropped and ground into a true cylinder, and sliced into thin wafers. Up to 60% of the silicon is wasted in the cutting and grinding and slicing.

Westinghouse has produced the web-dendrite technique. Two silicon filaments are dipped into molten-purified silicon and pulled up, forming a film resembling soap between the tines of a fork. This film is a single crystal of silicon.

In the edge-defined film-fed growth technique (EFG) developed by Mobil-Tyco, wider ribbons of polycrystalline silicon can be made by pulling the molten-pure silicon through a graphite die, using capillary action. Only 20% of the silicon is lost in this technique.

Honeywell developed a process that coats a thin ceramic strip with a layer of pure silicon by skimming the ceramic strip over a long, narrow trough of molten silicon. This process produces rectangular cells, as do the web-dendrite and EFG techniques.

The cells, once produced, are assembled into modules and wired in series to produce a standard output, normally 12-14 volts, for recharging conventional 12-volt storage batteries. The face of the module is then covered with lexan, glass, or soft, rubbery silicon for protection. The soft silicons, however, tend to accumulate dirt and do not protect the panel from damage if they are stepped on.

As you can see, the manufacture of a silicon cell is a tedius, expensive process, but the final result is the most fantastic generator you can imagine. It has no moving parts, operates on cloudy days – or even under water – as well as in the sun. It is silent, seldom wears out, uses free fuel, responds to light in 3 to 20 microseconds, and operates in temperatures ranging from -65 degrees C to +175 degrees C. Once it is installed you can forget about it. There are no costs for maintenance or operation. And its fuel belongs to all of us.

The majority of commercially available solar cells are made of silicon, but silicon is not the only material from which a photovoltaic cell can be made. Other types of cells are being studied in the effort to reduce costs, among them the cadmium sulfide cell. A cadmium sulfide cell is already available from SES, Inc. of Newark, Delaware. Research is being done in conjunction with the cadmium sulfide cell, on a cadmium stannate film to be used as a backwall with the cell for generating heat. However, the cadmium sulfide cell is less than half as efficient as its silicon counterpart at present.

Don't wait for the 15 cent peak watt before purchasing photovoltaics – first consider what they can do for you today. Among other projects, Energy Saving Systems of Holmes, Pennsylvania, is rebuilding a 45-foot sailing trimaran called *Free Energy*. This boat will be electric-powered for docking and motor sailing and the electricity will be furnished by photovoltaics. This same company operates many of their office machines from a solar array on the roof of their plant.

Six lighthouses on the Florida reefline from Sand Key at the southern tip of Florida northward to Fowery Rocks off Miami, have been tested and found economically more feasible when run by solar power and automation than when operated any other way. Methods tested were windmill-powered generators, thermoelectric generators, diesel generators, large battery banks hooked together, and low voltage, DC, Xenon flashtubes powered by solar-charged on-site batteries. The solar-charged system won hands down – the other systems proving to be either too unreliable or too expensive. The first solar panel was installed at Carysfort Reef light in 1975; the second at American Shoal light in March 1981. The lights on Fowery Rocks, Careysfort Reef and Sombrero Key are scheduled for completion by the end of fiscal year 1984. By the end of 1984, all six of the Florida Reef lights were to be powered by solar.

The Japanese have used sun-powered navigational aids since the 1960s. In addition to running navigational aids at sea and keeping battery banks charged, different sized solar cell arrays are being used right now to power satellites, radio transmitters to earth, weather stations, even wristwatches, flashlights and small fans. On an overcast day, a solar module will still generate enough power to operate a homing beacon forever. And if this is a marine module, it will withstand severe environmental conditions, most temperature extremes and total immersion in sea water with little loss of power output. What an insurance policy for blue-water sailors!

The only enemy of the silicon cell is excessive heat. Excessive heat reduces the efficiency of the cell and if it becomes *too* extreme, can cause the cell to crack. The face of the solar panel should be air tight to prevent a buildup of heat between the face and the cells. It should have an open back and be mounted in a slightly elevated position to permit easy dissipation of heat. If possible, it should be mounted so that the entire panel will swivel. This way, it can be tilted toward the sun on different tacks, and will simultaneously be ventilated from beneath.

When considering solar cell arrays for your boat, shop around. Different manufacturers offer different kinds and sizes of modules, as well as single cells. It is of the utmost importance in marine applications that the cells themselves be encapsulated and that the metals used in their manufacture be non-corrosive. The cells must also be protected from *extreme* buildup of heat. Also, prices are still competitive, so it pays to shop around. Some of the companies that supply solar cells, solar generators, kits and solar-operated marine equipment are listed below:

Ecotronic Laboratories, Inc.
7905 E. Greenway Rd.
Scottsdale, AZ 85260

Free Energy Systems, Inc.
Holmes Industrial Park
Price & Pine Sts.
Holmes, PA 19043

PDC Labs International
P.O. Box 603
El Segundo, CA 90245

Photocomm, Inc.
7745 E. Redfield Rd.
Scottsdale, AZ 85260

SES, Inc.
Tralee Industrial Park
Newark, DE 19711

Solarex Ventures Group
1301 Piccard Dr.
Rockville, MD 20850

Solar Power Corp.
Affiliate of Exxon Corp.
20 Cabot Rd.
Woburn, MA 01801

Solarwest Electric
(ARCO Solar panels)
232 Anacapa St.
Santa Barbara, CA 93108

Solenergy Corporation
171 Merrimac St.
Woburn, MA 01801

Spire Corp.
Patriots Park
Bedford, MA 01730

United Energy Corporation
1176-D Aster Ave.
Sunnyvale, CA 94086

Studying the specifications of the various solar modules can be very confusing. The output of photovoltaics is usually expressed in milliamperes and millivolts. Remember that one milliampere is only one thousandth of an ampere. Therefore, 500 milliamps is .5 amps or one-half an amp. It is also important to remember that the output of solar cells increases with increased light intensity, but decreases with increased temperatures. The values expressed in most solar module specifications represent the output of the unit at peak power conditions. Peak power ratings do not take into account losses for cable and blocking diodes, the time of day, clouds or shadows, or less-than-optimum panel tilt toward the sun. Therefore, a little research and some simple arithmetic will be required before selecting your solar panels. You will need to know the number of peak sun hours per day in your cruising area. Most manufacturers are happy to furnish this exact information, but you can roughly estimate a *yearly* average of 5 peak hours per day in U.S. cruising waters. The peak hours will vary from season to season, but will average about 5 peak hours on an annual basis in most places.

Now you are ready to select the proper solar array for your boat. Refer to your previously completed Electrical Consumption chart. Obtain the total from the "Daily Amp Hours Used" column. Divide this figure by 5 (the average peak hours) and multiply by 1.2 to compensate for the inevitable losses. This will give you the total ampere hours required.

$$\frac{\text{Daily Ampere Hours Used}}{\text{Yearly Average Peak Hours}} \times 1.2 = \text{Total Ampere Hours Required}$$

Now, divide the Total Ampere Hours Required by the output current (amperes) of the solar panel you are considering. Remember that a solar generator for charging 12 volt batteries should have a voltage rating of 14 to 14.5 volts, but the current (ampere) output will vary according to the size of the cells.

$$\frac{\text{Total Ampere Hrs. Required}}{\text{Ampere Rating of Panel}} = \text{Number of Panels Needed}$$

In practical terms, if you have equipment that uses an average of 60 Amp hours per day and you are considering a 14 volt solar panel that is rated at 6 amperes:

$$\frac{60 \text{ Amp Hours}}{5 \text{ Peak Hours}} \times 1.2 = 14.4, \quad \text{Then} \ldots$$

$$\frac{14.4}{6 \text{ (Panel Rating)}} = 2.4 \text{ or } 3 \text{ Panels}$$

Most cruising boats can easily accommodate 3 solar panels.

A voltage regulator is also recommended for all systems that produce over 20% excess energy to avoid overcharging the batteries, with the accompanying loss of electrolyte.

Selecting the appropriate battery bank for your solar panels is very important. The solar panel array is designed to provide all the energy you need over the year. The storage battery is the buffer between the solar panels and your electrical load, supplying power during periods of low sunlight or no sun, and accepting a charge from the array during periods of high sunlight. For this reason, its state of charge will vary both daily and seasonally. Therefore, the capacity of the bank of storage batteries must be chosen to meet these demands.

In addition, there will be times your solar panels are only trickle charging your batteries, which does tend to damage the structure of the standard lead-acid battery's plates. The plates of a deep-cycle battery are stronger than the standard 12 volt automobile battery and will stand up longer to the rigors of trickle charging. So, select the best quality, deep-cycle batteries you can afford for your battery bank and be sure they provide more than enough amperage to fill your 24-hour electrical needs. Match them to your panels, include a blocking diode, and add a voltage regulator if the solar array is capable of producing over 20% excess energy over the year.

If you find your electrical requirements are for a number of panels too large to accommodate, take another look at your Electrical Consumption chart with the idea of powering a limited number of items with photovoltaics – perhaps emergency equipment or low drain navigation equipment

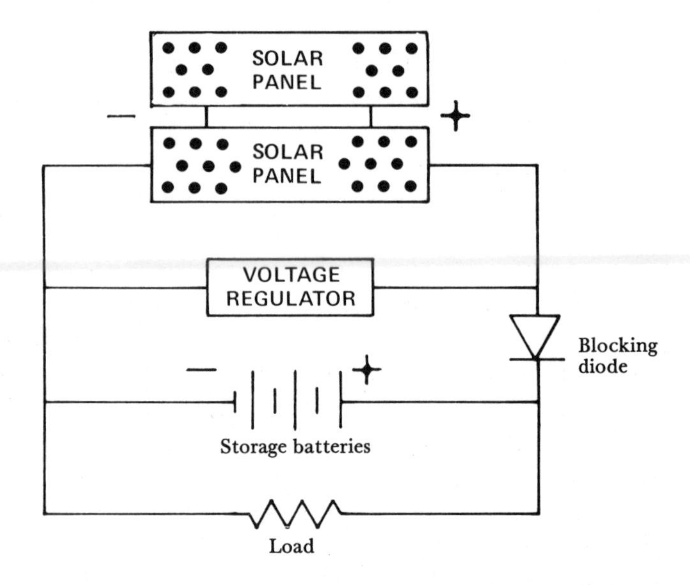

Complete Solar Electrical Generating System

such as knotmeters or depth-finders. In the Observer Single-handed Transatlantic Race that began June 5, 1976, off Plymouth, England, the most popular form of naturally-produced electrical energy for autopilots in the fleet of 125 starters was the solar cell array. The use of photovoltaics among offshore racers and cruisers has been increasing ever since.

Solar cells are so flexible, they can power equipment directly, can be used to recharge an isolated battery retained to power only certain equipment, or can be used as one unit in a larger generating system that incorporates other methods for producing power. Additional solar generators can be added at a later date, once you have proved their reliability to your own satisfaction.

When mounting a solar panel, it should be installed where it will receive the maximum amount of sunlight, such as on a cabin top. Most manufacturers recommend mounting the generators with an air space beneath the encapsulated panel, in order to ensure the proper dissipation of heat. If possible, mount panels so that they can be swiveled to face the sun more directly. This will increase their output tremendously.

It would be almost impossible to fracture a plastic-covered panel, but if the integral cover is made of glass, place it where it is least likely to be damaged by flying winch handles or other UFOs. Soft plastic-coated panels cannot be walked on. Lazarette lids are an out-of-the-way location favored by many blue water sailors for their solar panels but, to prevent excessive power loss, the panels should be located as close to the batteries as possible.

The solar panel will come with positive and negative lead wires that should be connected directly to the positive and negative terminals of your battery. If extra wires are spliced for added length, be sure the wire is heavy gauge, stranded, copper wire and that connections are well sealed and insulated. Be sure a blocking diode is provided to prevent battery discharge during hours of darkness. The panel itself is attached to your boat with adhesive, screws, or a swivel device, and the first ray of sunlight will activate your own private utility company.

Cabin Top Solar Cell Array

Some experiments are being done using concentrating collectors to gather additional sunlight and focus it onto the solar panel, in order to increase its output. But, since excessive temperatures decrease the efficiency of the cells, the whole panel must be cooled in order for this system to work well. This seems to be too complex a system for successful use on a boat.

Solar micro-generators are also available at reasonable cost, for recharging the many 1.5 volt rechargeable batteries that are used aboard most boats in flashlights, portable radios, tape recorders and pocket calculators. A solar flashlight is available with its own detachable solar cell array. It operates like an ordinary flashlight but keeps recharging in the sunlight to give you light when it is dark. Ready-made solar wristwatches, solar calculators, and solar anchor lights are also on the market. Small, reasonably priced solar arrays can be used to recharge *any* small *rechargeable* battery or to directly power a small, high-torque, low voltage, DC motor, such as a small exhaust fan or blower. The possibilities are limitless.

Another type of solar-powered generator consists of a conventional alternator or generator, turned by a solar-heated steam or hot air engine. Neither of these engines is practical for a small boat at the present time, but they both hold promise for the future.

The present day internal combustion engine is the result of trying to improve the problems inherent in the early heat engines. However, today, excellent steam engines in sizes suitable for generating electricity from 100 kilowatts to 10,000 kilowatts are sold by Spilling Consult AG, Sonneweg 4, CH-5610, Wohlen, Switzerland. Since these engines are designed for continuous operation, they are extremely heavy and unsuited for use on a *small* boat. However, several Spilling steam motors are in successful service at present as auxiliary engines aboard large, ocean-going ships. In addition, running a steam engine with solar radiation would require the use of a large focusing collector which would track the sun – a difficult design problem on a moving boat. Finally, a large area reserved for high quality heat storage would be required in order to run the engine when the sun is not shining.

Interest has been renewed in the Stirling engine (a heat engine) because these external combustion engines are so democratic about the type fuel they use. The Stirling cycle engine can use concentrated solar energy or it can burn kerosene, coal, straw, wood, sawdust, cardboard, discarded Christmas trees, or *anything* that burns.

A solar-powered Stirling engine produced electricity that could be fed into power company lines for the first time in August and November 1981 experiments at the Advanced Components Test Facility, part of Georgia Institute of Technology's Engineering Experimental Station in Atlanta. A 20 kilowatt external combustion engine (developed by United Stirling of Sweden) has pistons driven by solar-heated helium. A large mirror field (550 circular mirrors) concentrates sunlight on a Georgia-Tech-designed cavity receiver. The engine, operating at 750 degrees C, generated 440 volt-3 phase alternating current with 20% to 25% efficiency.

Several miniature working-model Stirling cycle engines and casting kits are available from Solar Engines, 2937 W. Indian School Rd., Phoenix, Arizona 85017.

Running a Stirling cycle engine with solar radiation aboard a moving boat would present the same difficulties as the steam engine – the problems in keeping a concentrating collector in focus, and the large heat storage area required. However, these engines are worthy of more study for use aboard ship, to operate not only generators, but also to operate fans, refrigerators or other equipment.

The wind generator also can be considered a solar generator. Wind is one of nature's solar energy storage systems. The sun heats the surface of the earth, causing the air to expand and build up a pressure gradient between one region and another. The pressure gradient is the storage system and the wind is the pressure relief valve. The main problem with using wind as an energy source is that it is randomly intermittent and therefore not totally predictable or dependable.

Wind Generator

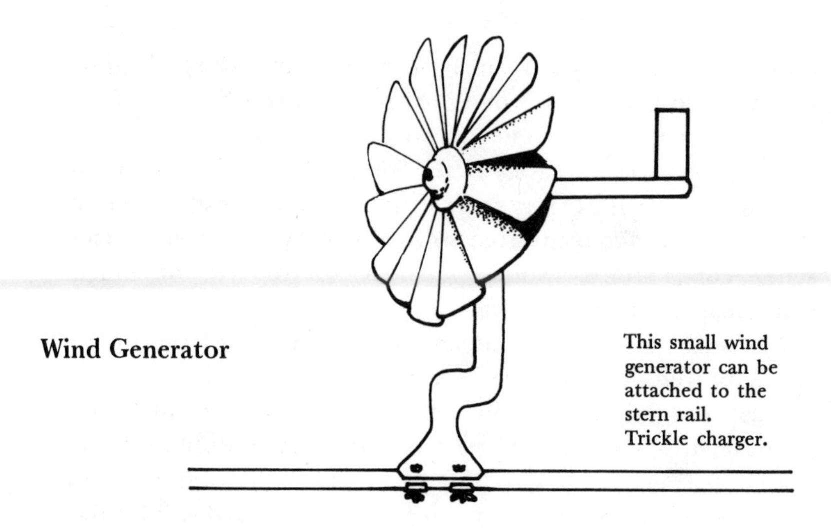

This small wind generator can be attached to the stern rail. Trickle charger.

Several commercially built wind generators can be purchased for use on your boat. They resemble the old fashioned windmill that Don Quixote jousted with. To be really correct, a windmill grinds grain, a wind pump pumps water, and a wind generator generates electricity. The blades of the wind generator are attached to a conventional alternator or generator which is connected to the battery bank. Most of these wind generators begin to charge at wind speeds of approximately 10 mph and are built to withstand gale force 8 winds without damage. The smallest commercially available wind generator weighs less than 10 lbs., is only 17 inches in diameter, and can be stored when not in use. This generator is often mounted outboard on the spreaders of a sailboat or clamped to the bow pulpit or stern rail.

Large wind generators are available with longer blades – up to 9 feet. These generators are correspondingly more expensive, but have a much higher electrical output. The main problem with these large generating plants is finding an aesthetically pleasing place to mount them. One possibility is on a spar mounted on the stern of a sloop, or (if you have sufficient clearance) atop the mizzen of a yawl or ketch. Owners of large power craft have more flexibility in location.

Electrical output of a wind generator varies from milliamps for the smallest up to 50 amps for the models with a 9-foot blade.

There are two major disadvantages to wind generators. One is that they are extremely noisy at high speeds. They should never be mounted directly over sleeping quarters. The other problem is that the wind is a sometimes thing, much less predictable than sunshine, and, since the rate of charge depends on the wind velocity, there is no way to totally depend on the wind to supply your electrical needs.

When purchasing a wind generator, be sure to study the manufacturer's specifications carefully. There could be a vast difference between having the *capability* of keeping batteries fully charged and doing so. A lot of "ifs" are involved: *if* the wind is blowing, *if* it blows long enough, *if* it blows hard enough. The capacity of the battery is another limiting factor. However, the wind generator is an excellent auxiliary to a more elaborate electrical generating system.

Wind generators are available through the following companies:

AMPAIR Products
Aston House, Blackheath
Guildford, Surrey
England GU48RD

Eodyn Wind Charger
Viscom International Inc.
244 Farms Village Rd.
West Simsbury, CT 06092

Hamilton Y. Ferris
Box 129
Dover, MA 02030

Greenwich Power Log
9507 Burwell Rd.
Nokesville, VA 22123

Kucharik Wind Electric
P.O. Box 786
Toms Rivers, NJ 08753

Mini Windtap
ESSCOA Mfg. Div.
P.O. Box 318
Newfoundland, NJ 07435

Web Charger
Old Ships Way
Box 586
Provincetown, MA 02657

If you prefer to build your own wind generator, "Wind and Windspinners" by Michael A. Hackelman and David W. Hause is an excellent manual on the basics of wind-to-electric conversion. It is available from Mother's Bookshelf, P.O. Box 70, Hendersonville, North Carolina 28739. Also, Windlight Workshop, Rt. 2, Box 271, Santa Fe, New Mexico 87501, will provide a catalog of parts, props, controls and motors for a small fee.

Another natural auxiliary to solar and wind generators is the hydro or water generator. The water generator consists of a conventional type alternator/generator that is driven by a spinner trolled through the water behind the boat. Several commercially built water generators are sold, but they vary widely in design. While one model is self-contained and mounts on the stern of the boat in an outboard motor bracket, another model uses a spinner connected to its remote generator by a torque line. They all depend on a generator that begins to deliver electrical power at relatively low rpm's. How much current is delivered depends on the boat's speed through the water, but typically will be from 5 to 16 amps at 12 to 14 volts. For more information on commercially built water generators, write the following distributors:

"AQAIR 50"
AMPAIR Products
Aston House, Blackheath
Guildford, Surrey
England GU48RD

Hydrocharger
Regent Marine & Instruments, Inc.
1051-B Clinton
Buffalo, NY 14206

Hamilton Y. Ferris
Box 129
Dover, MA 02030

Hydroalternator
Motorola Marine
1313 E. Algonquin Rd.
Schaumburg, IL 60196

Vega-C Research
P.O. Box 568
Miami, FL 33133

Another way to use your boat's motion through the water to generate electricity is to belt a high-performance type generator directly to your free-wheeling propeller shaft. You must have the type of transmission that allows free wheeling of the shaft without overheating or other damage to the engine. If your engine does not have this type of transmission, write the following company for full information on their "Velvet Drive" gears:

Warner Gear
Division of Borg-Warner Corp.
P.O. Box 2688
Muncie, IN 47302

Free wheeling under sail or at trolling speeds with one or more engines shut down saves fuel and reduces drag, in addition to providing the option of running a generator off the propeller shaft.

When making an installation of this kind, it would be best to work with a competent marine electrician to avoid any possibility of damage to your engine, battery bank, or electrical circuits. Even though this is a simple and fairly inexpensive modification, each engine installation is too individual to be able to cover all possible problem areas in this text. It is important to remember that it will be necessary to install another voltage regulator when adding a conventional alternator/generator. Trickle chargers, on the other hand, unless they are capable of exceeding your battery bank's capacity by 20%, do not require a voltage regulator and need only a blocking diode and a quick disconnect plug.

If, after improving your present generating system to include one or more of the natural free-fuel generators, your electrical demands still outstrip your ability to meet them, the next step is to reduce these demands.

4
Reducing Your Electrical Demands

Study your Electrical Consumption chart carefully again, this time with thoughts of reducing or eliminating as many electrical demands as possible.

First, isolate high amperage items, just to become familiar with which ones they are. High on the list will be engine start, all pumps, lighting – particularly spreader lights and spotlights – and refrigeration. We should consider non-electric alternates to as many of these items as possible. Substitutes will not be acceptable for safety equipment, such as engine start and bilge pumps, even though their drain on the electrical system is high. Therefore, it is best to provide a separate, isolated 12-volt battery for these safety devices. Manual alternates should also be provided.

The high draw of your lighting system may come as a distinct surprise to you, especially when you already are having trouble reading by the available cabin light at night. But never fear, there are many alternate possibilities to be considered in cabin lighting.

Natural solar lighting should be increased in every possible way. Add as many small, opening portlights as you can. They should be small in order to withstand heavy seas, and they should open in order to provide a flexible ventilating system. There will be space on some boats to add opening hatches. Hatch covers, if not already covered with photovoltaic cell arrays, either can be made from a transparent material, or a small "window" or transparent ventilator can be installed in the center of each hatch cover. Cabin doors can be louvered to admit more light, in addition to more air, provided that heavy storm doors also are provided for protection from raging seas and cold weather.

Deadlights are another excellent way of increasing the amount of natural light that finds its way below. They are usually round, 4 to 8 inches in diameter, or rectangular, 6 to 10 inches long, made from glass or plastic, and mounted flush with the deck. Search the marine specialty shops and salvage yards for a few of the old fashioned prism-type deadlights. These are superior in every way because they are made from very heavy glass and the prisms provide more light-gathering surface than flat glass. They let more light in and are prac-

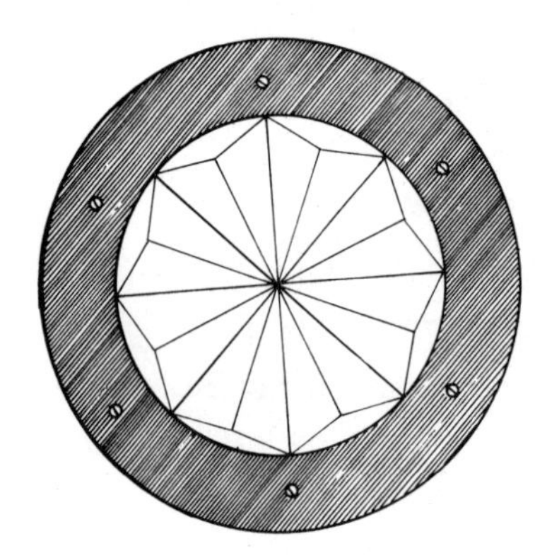

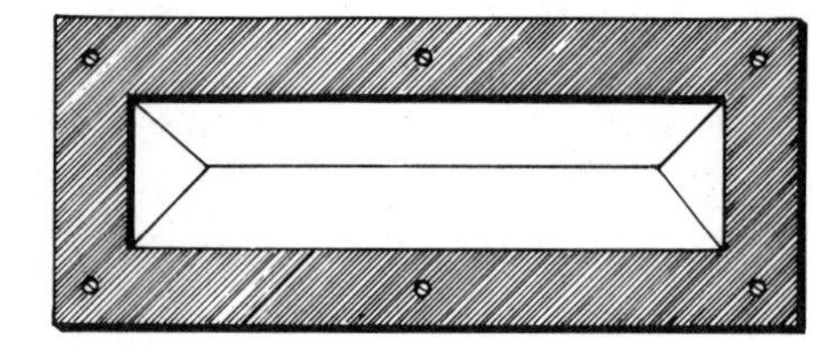

Old-Fashioned Prism Type Deadlights

tically unbreakable. All deadlights must be carefully mounted, bedded and sealed or they are a potential source of leaks. Try a deadlight over your hanging locker if you want a feeling of real luxury as you thumb through your nautical wardrobe.

In their book *Living Aboard,* Jan and Bill Moeller suggest a unique way of increasing natural light below deck. They suggest replacing the solid tops of Dorade boxes with clear or frosted plastic.[1] Even if you have molded fiberglass Dorade boxes, the tops can be cut out with a small electric saber saw, and new, clear plastic tops can be screwed in place with self-tapping screws. One of the new, tungsten-carbide saw blades will make child's play of this job. These blades are available at any large hardware store.

Anything you can do to bring more natural light into your cabins will reduce the amount of artificial light you will require. The Moellers claim that the clear plastic-topped Dorade box provides as much light as a deadlight.

When artificial light is needed below deck, fluorescent lights should be selected rather than the incandescent lights that are standard on most of today's boats. Fluorescent lights put less drain on the battery while providing more light. Since fluorescent fixtures are long and narrow, they are perfect for installation under cabinets and lockers and over work areas. (Many people consider them too harsh for reading.)

To completely eliminate the electrical drain for lighting below deck, try the humble kerosene lamp. Although it gives off some heat, the kerosene lamp is a pleasant addition to any boat, and it is versatile enough to burn most any kind of oil, not just kerosene. Many beautiful styles of wick-type oil lamps are available, but remember, when selecting one, that the amount of light produced is in direct proportion to the width of the wick. Also, remember that wicks have to be replaced regularly and chimneys occasionally. For these reasons, a sturdy hardware store oil lamp might be a more sensible selection than one of the elegant, traditional, gimbaled brass marine oil lamps. The marine lamps are usually quite small, with correspondingly narrow wicks and the cost of replacing their wicks and chimneys is usually astronomical. Spare wicks and chimneys for the common hardware store oil lamp will be inexpensive and universally available.

For the brightest, cleanest light from a wick-type oil lamp, you must regularly wash the chimney, top the tank with oil and trim the wick. This is a task that should be included on the regular duty roster. If the oil lamp you select is not gimbaled, be sure that it has either a wall mounting bracket and overhead smoke bell, or a handle. Then it can be hung on a hook in any convenient place or stowed when not in use. Hang the lamp in front of a mirror or brightly polished piece of metal for increased light.

The most efficient oil lamp uses a mantle rather than a wick and may operate under pressure. This type lamp will

Oil-Burning Cabin Lights

give off as much light as a 100-watt bulb, while using very little fuel. These lamps are quite large, but they give off more light than several of the smaller wick-type oil lamps. Shades are often available for these lamps, which diffuse the intense light, and turn them into excellent reading lights.

For safety reasons, gasoline pressure lamps should never be used aboard ship. Propane lamps, although relatively safe, are too expensive to operate on a regular basis. Individual battery-powered lamps are sold which save your ship's main battery bank, but these lamps wear their own batteries out very quickly. Unless you have a solar array specifically designed to recharge battery-powered lamps, they cannot be considered a dependable regular light source.

An energy-efficient offshore cruiser will be equipped with both fluorescent electric cabin lights and oil lamps, as well as a handful of candles to use when all else fails.

Lighting above deck is another story. Spreader lights and spotlights should be considered emergency equipment. When tending sails at night on a wildly heaving foredeck in mid-ocean, nothing takes the place of well-aimed spreader lights. The same applies to the spotlight when trying to pick up a man overboard or locate a difficult marker. Even though these situations are not daily occurrences, it is best to allow enough amperage to take care of them. If time permits during emergency situations, the engine can be started to provide electrical power while these lights are operating.

Standard electric running lights can be replaced with oil-burning running lamps. These lamps are available in brass, copper or galvanized metal, in several sizes, through most marine supply houses. A secure mount must be provided to keep them from slipping off their brackets. (These are ordinarily attached to name boards or ratlines mounted on the shrouds.) They are usually wick-type oil lamps and should be cared for exactly like the wick-type oil burning cabin lamps. A regular schedule for filling, cleaning and wick trimming is essential. These lamps usually function quite well in heavy winds, but have been known to blow out. For this reason, it is important to keep your electric running lights in good operating condition in case of emergency.

Most cruising boats use an anchor light powered by its own battery. These low intensity anchor lights are very energy efficient and will operate for up to two years on one standard 1½-volt ignition dry cell battery. If you are using a less efficient anchor light, consider either changing to this type light or to an oil burning anchor lamp to eliminate another demand on your ship's battery bank.

The next item with a big electrical appetite is the pump.

The electric bilge pump, as mentioned earlier, can be considered emergency equipment, but you may have one or more additional pumps.

A small sump is often provided into which water from a shower or from an icebox drains. A pump then lifts water from the sump and discharges it overboard. Instead of using a separate pump for this purpose, it would be more energy efficient to gravity drain the sumps directly into the bilge and use the bilge pump to discharge this water overboard. Icebox and shower drains should be screened in this case to keep foreign material out of the bilge. Soapy water from the shower should keep down any unpleasant odors that might find their way into the bilge from the icebox. Or, pour a strong solution of water and baking soda down the icebox drain once a month, just to be sure. You will probably have the cleanest bilge in the marina.

A manual switch should be provided for the electric bilge pump in addition to an automatic switch. The manual switch can be used when you are aboard to turn the pump on only when the bilge is really full, and the automatic switch can keep an eye on the water level when you aren't there to do it for yourself.

Another power hog can be the pressure water system pump, a totally unnecessary piece of equipment, but the last electrical appliance most people will voluntarily give up. This little pump can make the difference between *living aboard* ship and *camping out* in an endurance contest. So, rather than eliminating this piece of equipment, I would favor using it with discretion, and adding a header tank and sink sprayer to reduce its electrical demands. A header tank is simply a hollow air chamber into which a small amount of water is pumped against air pressure. This maintains a reserve, which will reduce the amount of cycling your pump must do. The header tank should be mounted vertically and as close to the pump as possible. Header tanks are available commercially from marine plumbing supply outlets, or can be constructed very easily by the home craftsperson.

A galley sink sprayer also helps conserve electricity and water because it provides a fine but forceful spray while using a minimum of water, again reducing pump cycling time. *Thoughtful use* of an efficiently designed pressure water system can be considered "elegant frugality." But, even with an energy efficient pressure water system, complete with sink sprayer(s) and header tank(s), a manual backup is a necessity.

The only thing worse than not having water is having it and not being able to get it out of the tank. The manual backup also gives you another option when sterner electrical conservation measures are required.

If you do not have a pressure water system aboard your boat, consider installing a gravity feed system instead. Heavy water tanks usually are installed as low as possible in the bilge to maintain the boat's stability. A small auxiliary tank could be added atop the cabin or close to the overhead inside the boat. Water could then be pumped from the main tank by hand, or poured into the auxiliary tank daily (or as needed), to gravity feed the faucets. Remember that 231 cubic inches contain one gallon of water, and it will be easy for you to design your own tank to fit your allotted space. For instance, if you want a 5-gallon tank (231 cubic inches × 5), you will need a tank that contains 1155 cubic inches. Decide on the depth you prefer; since a low profile is usually desirable, let's use 2 inches as an arbitrary depth. Divide the size of the desired tank (1155 cubic inches) by the depth (2 inches). The result is 577 square inches, which represents the width times the length of your proposed tank.

Study your available space, then try different widths to determine which shape tank you prefer, by dividing 577 inches by the width. You will find that a 5-gallon tank can be 12 × 48 inches, or 24 × 24 inches, or 18 × 32 inches, or any one of many combinations. Experiment until you find the shape that is exactly right for your boat.

The following companies either stock water tanks in standard sizes or will build a tank to your specifications:

W. H. Denouden, Inc.
Box 8712
Baltimore, MD 21240

Richmar Industries
6900 NW 37th St.
Miami, FL 33147

KRACOR, Inc.
1043 13th Ave.
Grafton, WI 53024

Wilcox-Crittonden
699 Middle St.
Middleton, CT 06457

Or check your telephone directory under "Tanks—Custom."

Be sure your tank is fitted with an air vent, as well as fill and discharge pipes. A large, clean-out, access plate is a desirable addition. The finished tank should be installed in the most convenient place above faucet level, and attached to the boat with heavy duty mounting brackets or heavy strapping to prevent any possibility of slipping. A foot pump, rather than a hand pump, makes transfer of water from the main tank to the auxiliary tank a quick, easy task. A check valve in the fill hose will prevent water from siphoning back into the main tank, and the discharge hose can be attached to any high quality, chrome over brass, home-type faucet. A sketch of an auxiliary water tank is shown at right.

Continue to study the items on your Electrical Consumption chart, one by one, considering the different substitutes you might use and whether or not you could live comfortably with these substitutes.

An item which will come under immediate fire will be the electric marine head. The controversy about marine sanitation devices is still raging, and the laws governing these devices seem unenforceable. The available Coast Guard-approved sanitation devices waste electricity and are far less serviceable than the old fashioned hand-pumped head that discharges directly overboard. Many yachtsmen have removed their marine toilets, sealed the thru-hulls, and installed heavy duty buckets in their place. This may seem like a harsh solution to the problem, but, from an ecological point of view, it would take an awful lot of yachtsmen discharging waste overboard to equal the waste one whale or other large sea creature puts in the water daily. And, so far, I know of no government regulation that requires stoppers for whales.

Regardless of the controversy, your problem is how to comply with the law in the most energy efficient way possible. Only you can decide which marine sanitation device you can live with, after considering their good and bad points.

Flexible holding tanks are available that can put inaccessible areas to work for this undesirable task, but if the tank should rupture under stress, cleanup would be impossible. All holding tanks occupy space, fill up when you're not looking,

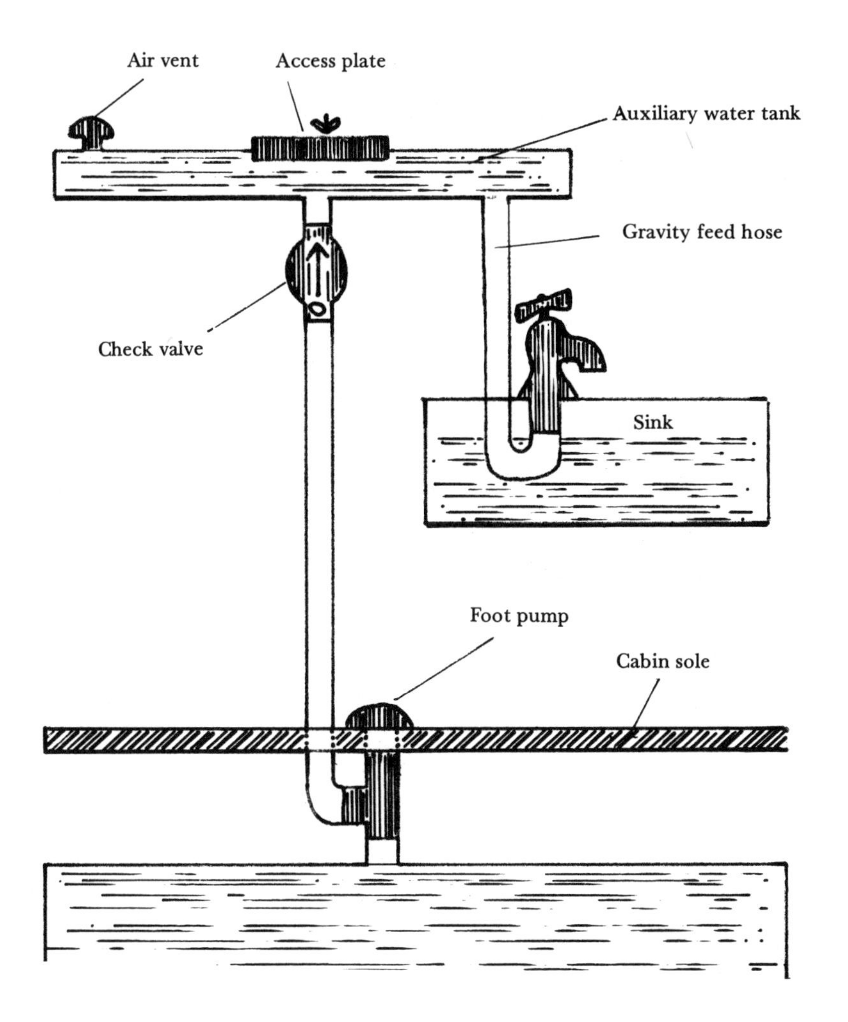

Main Water Tank

and are heavy when full. They have been known to spill their contents in heavy seas. Pump-out facilities are not readily available in many areas, and if the contents are pumped overboard (legally or illegally), the chemicals used for odor control rapidly disintegrate bronze seacocks and thru-hull fittings. Even with the addition of chemicals, holding tanks can generate foul odors, and sometimes even explosive methane. This happens when anaerobic decomposition takes place.

There are two kinds of organic decomposition: *anaerobic* (in the absence of oxygen) and *aerobic* (requiring oxygen). Any kind of organic animal or vegetable material can be broken down by either process, but the end products will be very different.

Anaerobic decomposition takes place in two stages. Acid-producing bacteria must first break the organic molecules down into simpler sugars, alcohol, glycerol and peptides. Once these substances have accumulated *in sufficient quantity,* a second group of bacteria convert some of the simpler molecules into *methane* gas. Traces of other gases, carbon dioxide, hydrogen, a small amount of heat, and quantities of slurry are also produced.

Holding tanks could become useful and attractive *if* someone would turn a holding tank into an anaerobic digester to *deliberately generate methane gas;* then devise a method to capture and use this gas aboard ship to operate a stove, refrigerator or engine; and finally obtain Coast Guard approval. But, space limitations seem to make an anaerobic digester tank for a small boat an impossible engineering achievement at present.

The recirculating toilet employs a holding tank of a sort. These toilets are "charged" with several gallons of water and odor control chemicals. They recycle this water each time the toilet is flushed, until its self-contained holding tank is full; then it must be emptied somewhere. A pump-out facility can be used, or the portable models can be carried to an onshore toilet and emptied there. They are inexpensive and are available in manual or electric models. However, my personal opinion is that they are too bulky, taking up valuable space in a small area, that they fill up faster than anyone dreamed possible, and that they usually smell bad in spite of the addition of expensive chemicals. There is also something basically offensive to me about carrying about your own waste, recirculating it, and pumping out great quantities of it. In spite of their undesirable qualities, though, the portable, manually-operated recirculating toilets are energy efficient. For this reason, one of these heads might be the right one for you.

There are several Type 1 flow-through sanitation systems on the market at the present time. These devices macerate and decontaminate waste before discharging it overboard. If you are seriously considering this type device, you should request brochures from the manufacturers and make your own comparisons. Each device has something different to recommend it, and only you can decide which sanitation system is right for your boat. For instance, the LECTRA/SAN requires no chemicals when operated in sea water and only table salt when used in fresh or brackish water. The IWSS System 1000 marine sanitation device is a small, completely self-contained unit. The Mansfield TDX requires power only when its small tank is full, instead of after each flush. Write the following manufacturers for information on their Type I sanitation devices:

Galleymaid Marine Products
Box 10417
Riviera Beach, FL 33404

Mansfield Sanitary, Inc.
150 First St.
Perrysville, OH 44864

International Water Saving System
387 Granite St.
Braintree, ME 02184

Raritan Engineering Co.
1025 North High St.
Millville, NJ 08332

A Type II flow-through marine sanitation device which meets Coast Guard requirements is now available in a size appropriate for most recreational boats. This unit is called the MICROPHOR and can be purchased from Southern Marine Sanitation, Inc., 1650 W. Oakland Park Blvd., Fort Lauderdale, Florida 33311. Write them for more information. This unit consists of an *aerobic* digester tank and a chlorinator which can be connected to any low-water-usage-type marine head.

An aerobic digester produces carbon dioxide, ammonia, small amounts of other gases, considerable heat and a small amount of dry residue. With this particular unit, wastes are pumped into the digester tank, where solid wastes are broken down by aerobic decomposition. Liquids drain from the digester tank by gravity flow, through a chlorinator, then overboard. The bottom of the digester tank must be above the water line in order to gravity flow the effluent overboard.

A composting toilet is also available that has no overboard discharge to worry about. Waste is transformed aerobically into earth-like humus. A year of daily normal use by a family of five or six will produce only enough odorless, powder-dry residue to fill a coffee can. This residue collects in a tray at the bottom of the unit and should need to be emptied only once each year. This type toilet has an electric heating coil and a small fan to speed up the digesting process and evaporate liquids. It is also quite heavy, but there is much to recommend it. You may be able to make certain alterations to reduce its electrical demands. Venting through a tall stack, such as a hollow aluminum mast, with a smaller solar-powered fan, could reduce energy consumption. This heating coil would operate less in a warm climate than in a cold one, or you may be able to use waste heat from some other source to further reduce its electrical demand. If you are interested in learning more about this type head, write to Recreational Ecology Conservation of United States, Inc., 9800 W. Bluemound Rd., Wauwatosa, Wisconsin 53226, for a complete descriptive brochure.

Your ultimate decision on the type head you install will depend on many things: the space you have available; whether you use your boat for racing or cruising; your electrical capacity; how often you have guests aboard (and how many of them there are); your waste handling tolerance level; and your finances. It's an important decision, so take your time. This is one function you can't issue ration cards for.

As you reduce or eliminate your electrical demands one by one, change your Electrical Consumption chart to reflect these changes. New levels of independence will begin to come into view. At this point, it is quite possible that the only thing keeping you at all dependent on fossil fuels is your refrigerator. The marine refrigerator is the biggest energy user of all aboard most boats, and the one we most loath parting with. But there are alternates—even for this important piece of equipment.

5
Marine Refrigeration Solid State to the Icy Ball

Marine refrigeration – the bane of the blue water sailor – is a complicated maze of tubes and wires and imagination with a propensity for unexpectedly breaking the chain of command. It is something our hardier ancestors would have scoffed at, and some sailing purists are still scoffing. Yet, what piece of equipment adds more to the good life at sea? And what piece of equipment contributes less to your peace of mind?

There are many different types of marine refrigeration units available today, all of them hanging on to some gasping power source by a tenuous thread. The kind of cooling device you select can be determined only by analyzing what a refrigerator means to you. Is it a machine to make ice and cool beer? Or is it some place to store a food you can't live without, such as chocolate ice cream or salad greens? Is the luxury of refrigeration valuable enough to suffer the deafening roar of a generator breaking the silence of a moonlit cove? Is it important enough for you to rearrange your cruising schedule around necessary and sometimes unexpected pit stops for battery recharging and refrigeration maintenance? Only you can decide.

Let's examine each type and consider some of their advantages and disadvantages.

Solid State Icebox Conversion Kits. These new units operate on an old principle – the Peltier effect. When voltage is applied across two dissimilar metals, heat is absorbed from one metal and transferred to the other. The small, light-weight modules that employ this principle can be installed in any well-insulated icebox (up to 5 cubic feet), and are available in 12-volt DC and dual 12V DC/110V AC models. They are easy to install, inexpensive, and have low power drain compared to other types of refrigeration, provided the ambient temperature is no more than 75 degrees F.

The main disadvantage to these units is that they do not make ice. Also, if the ambient temperature is above 75 degrees F, the box requires additional insulation and a higher power drain can be expected, making it less than satisfactory in the tropics. If the box is larger than 5 cubic feet, additional modules must be added.[1]

Custom-Built and Conversion Units. Many firms offer refrigeration components that can be built into your boat wherever space permits. The evaporator (freezing compartment) or holding plates can be installed in an existing icebox or in any well-insulated container, and the compressor/condenser can be remotely located as far as 12 feet away if desired.[2] These components can be purchased as single voltage units (12V DC or 115V AC), dual voltage units (12V DC/115V AC), engine drive, and combination engine drive/electric single or dual voltage units. The refrigerator that uses an evaporator requires a nearly constant source of power.

Holding plates, on the other hand, contain a eutectic solution and, once frozen, hold for an extended period of time without power (12-48 hours). The relative efficiency and power demand of these units depends on the size of the insulated box, how well the box is insulated, and the ambient temperature, as well as the innate efficiency of the unit itself. These units make excellent use of available space, but are quite expensive to install, use and maintain.

Ready-Made Refrigerators. There are many small home refrigerators on the market, as well as a large number of small refrigerators built especially for boats, campers or planes. The household-type 115V AC refrigerators can be operated dockside or with a generator when offshore. The specially-built refrigerators are manufactured in single or dual voltage, propane, and some three-way models are sold which operate on 12V DC/115V AC/Propane.

These ready-mades are the least expensive type of refrigeration units to purchase. They are easy to install or remove for servicing, and well known brands offer nationwide service. But, although some top-loading models are made, most of these units open from the front, which results in spilled food if opened in rough seas or on the wrong tack. All front opening boxes spill volumes of cold air every time they are opened. Also, they are poorly sealed and insulated. All of these factors cause them to use power in enormous gulps.

Most boat owners in this country will not even consider having a refrigerator on board that uses an open flame, although this type unit is common aboard European boats. A good, outside-vented location and proper handling could overcome any feeling that propane is unsafe.

If you use a conventional mechanical refrigerator, you can count on running your engine at least two hours a day to support refrigeration equipment alone (excluding other power demands), whether it is to operate a generator, recharge batteries or turn an engine-driven compressor.

Iceboxes. Iceboxes of some description are built into most boats today. They may be tiny storage compartments or giant reefers that hold 300 lbs. of ice and they can be the most efficient, energy-free equipment on board. How well your icebox functions will depend on four things: the ambient temperature; the quality of the ice; the thickness and quality of the insulation and door seal; and you.

Be sure the icebox is protected from the direct sun and from exhaust heat from the engine compartment. Use fresh, block ice. It costs the same as cloudy, half-melted blocks, less than cubes, and it holds longer. (Dry ice is never available in marinas, and you can't put it in your Island Punch, anyway.)

Be certain the lid or door makes a complete seal when closed. Add heavy duty rubber weatherstripping, if needed. Open the box as infrequently as you can. Keep the box as full as possible. Fill empty spaces with ice, small cans of juice, soft drinks, beer, or whatever you have. But keep it full. Many commercial fishermen pack all empty spaces with crushed ice.

Add insulation, if necessary. Even seemingly inaccessible ice chests can be better insulated by removing the top or cutting a hole near the top of the box and pouring in liquid polyurethane foam. (Be sure to stop up drains or limber holes to avoid a bilge full of foam.)

Tevake was built in the northeast, and her huge, 200-lb. ice chest, while functional around the Boston area, was next to useless in the tropics. It gulped down 50 to 100 lbs. of ice a day. The addition of an awning, which shaded the icebox, and two gallons of polyurethane foam insulation poured around the built-in chest, reduced her ice consumption to 25 lbs. every second or third day.

Portable ice chests are a good investment, too. They can be loaded at home for day sails, or accept the overflow from other cooling devices when necessary. The lightweight, inexpensive plastic chests that are readily available in many sizes and qualities can be a good choice. But, if you are serious about keeping things cold for long periods of time, especially in tropical climates, look for one of the heavy duty, "commercial" ice chests. They cost more initially, but their superior insulation will save you money in the long run.

The main disadvantage of an ice chest is the cost, availability and transportation of the ice. If you have ever carried a 25-lb. block of ice three miles on your bicycle, then dropped it while trying to board . . . if you have ever cruised the Bahamas, gone eight miles off course, with a rapidly deteriorating box full of food, only to find that the guy ahead of you just bought the last misshapen piece of cloudy, brackish ice . . . if you live aboard and just added up your monthly ice bill, you will wonder if an icebox is really the answer.

The Cool Locker. Doing without refrigeration does not necessarily mean suffering and deprivation. I was puzzled by the many "ice" boxes without drains I saw in private homes in New Zealand a few years ago, until I was told (in a manner suggesting that I must be a bit balmy) that these were "cool lockers," ice being a totally unnecessary commodity. In these small boxes were kept dairy products, eggs, cheese wrapped in vinegar-soaked cheesecloth, and other items I had previously considered perishable. This brought me around to a whole new way of thinking about what sorts of things needed to be refrigerated.

At this point, *Tevake's* engine had been running over three hours *every day*, just to support her latest "improvement," a tiny refrigerator. A quick tour of several friendly, neighborhood marine refrigerators (my own included) revealed that they most often contained: several varieties of jams and jellies, left-overs, eggs, margarine, left-overs, milk, fresh vegetables, left-overs, a box of baking soda, left-overs, catsup, mustard, pickles, left-overs and assorted cold drinks.

I suddenly realized that few, if any, of those things really needed refrigeration. If we just ate or threw away our left-overs (or tried harder not to manufacture them in the first place), instead of storing them in the refrigerator for great periods of time, only to throw them away later, everything else on the list could be transferred safely to a cool locker.

There are as many ways of storing eggs as there are people storing them, and most methods work. Eggs keep the longest if you start out with fresh, unwashed, unrefrigerated eggs, stored in cartons, small end down, in a cool area such as the bilge. Unfortunately, unless you are on a first name basis with a chicken farmer, you will probably have to settle for the well-bathed, refrigerated supermarket variety. Even these eggs, if fresh, will keep in a cool place up to four weeks in the tropics, if you will bring them slowly to room temperature, then seal them any way you choose (waterglass, oil, wax, grease) and store, as above.

Milk can be made fresh daily from dry milk powder. If it is to be drunk, mix it several hours ahead of time, pour into a water-tight container and hang it overboard to cool. Sterilized whole milk in cans or cartons is available in some parts of the world. I have seen it sold in remote areas of the South Pacific, and in a few of the stores in this country that specialize in dehydrated and freeze-dried foods.

Many fruits and vegetables keep well for long periods of time without refrigeration. They keep best if never refrigerated, but if brought slowly to room temperature, then hung in mesh bags to ensure good air circulation, many of them will keep for long periods. All citrus fruits, apples in season, green tomatoes, and green bananas keep well. Eggplant will keep at least a week without chilling. Cabbage, sweet potatoes, and squash keep a month or more. Fruits and vegetables can be successfully dried and stored in a cool place for extended periods if desired. See Chapter 9.

Many well-cured cheeses will keep for months if wrapped in vinegar-soaked cheesecloth to reduce formation of unwanted mould. Your supermarket dairy manager can tell you which cheeses will and will not keep well. Processed cheeses need not be refrigerated, but their life is limited once they are opened, so buy in small sizes.

Butter and margarine are available whole or powdered in cans. Janet Groene, in *The Galley Book,* states that regular supermarket margarine can be brought to room temperature, softened, stored in clean, scalded containers, and kept safely for several months.[3]

Fresh meat and fish *will not keep* in your cool locker. This is no problem for the vegetarian, but the meat-and-potatoes yachtsman must buy or catch only as much as can be used at one time. You can also use the many fish and meat products available dried or in cans at the local supermarket. Or dry your own. See Chapter 9.

Kick the cold drink habit if you can. Some feel it is bad for your health, and I can personally attest to the fact that ice is habit forming. You will feel noticeably cooler in hot weather if you don't succumb to icy food or drink. Drinks can be *cooled* by placing them in a mesh bag and hanging them over the side. The deeper the bag hangs in the water, the cooler your drinks will be. Salad greens could benefit from the same treatment. Double-bag them in water-tight plastic bags, or place them in any water-tight container. Put the container in a mesh bag and hang it over the side. Remember, especially in the tropics where the surf temperature can be 80 degrees F, to hang the bag as deep as possible – below the thermocline, if you can.

Don McCoun tells us how to be really inventive with his recipe for ice-less ice. Place the item to be frozen or chilled in an earthen jar. Place this jar in another earthen jar. The second jar should be large enough to leave a space between the walls and bottom for the following mixture:

10 oz. Sal Ammoniac
10 oz. Salt Petre
1 lb. Glauber Salts

Add one quart of water and pour mixture into the larger jar. Cover both jars until item is frozen or as long as you want the item chilled.

DANGER: This is a highly corrosive mixture. Do NOT use tin vessels; use crocks or jars instead. Do NOT allow mixture to touch the item to be chilled, or your eyes or skin.[4]

The Solar Refrigerator. If you don't want your cruising schedule to be ruled by your need for ice, dockside electricity, fuel to run the generator, repairs, or constant restocking of the larder, consider the solar refrigerator. A commercial model will be available in the near future, but meanwhile, this can be a worthwhile do-it-yourself project. You could be the first one in your area to use this new-old device aboard your boat.

Commercially-sold refrigerators are either compressor systems or heat absorption units. In an absorption unit, a chemical mixture is circulated by heating (with flame or electrical element). The circulating mixture enters the evaporator inside the refrigerator, where heat is removed and transferred to a condenser outside the box. It would seem at first glance that a refrigerator of this type could be adapted to solar use if it could be located in a position where a Fresnel lens or parabolic reflector could be focused on the generator at the lower back of the unit to furnish heat in lieu of a gas flame or electrical heating element. However, the lens or reflector would have to be designed to track the sun in order to best use its heat, and this would be an almost insurmountable problem on most boats when underway.

An *intermittent* heat-absorption-type refrigerator, such as Crosley's "Icyball," would be more suitable for use on a boat. This refrigerator was manufactured and sold in Canada and the United States in the 1920s and 1930s. The patents for Crosley's Icyball have long since expired, but a few of these ingenious refrigerators may still be resting, long forgotten, in someone's attic or barn. A condensed version of Crosley's original "Instructions for Crosley Icyball Refrigerator" begins on page 62.

The Icyball consisted of nothing more than a top-loading insulated box with a detachable refrigerating mechanism. The refrigeration mechanism consisted of two balls connected by tubing, through which an ammonia-water mixture was circulated. The generator end of the mechanism was placed over a heat source (a tiny kerosene stove was furnished) and the other end was placed in a bucket of cool water. After heating the generator for 90 minutes, it was removed from the flame and the mechanism was placed in the insulated cabinet with the generator on the outside and the freezer end inside the chest. The cold end of this mechanism not only made ice cubes, but also refrigerated the box for 24 to 36 hours.[5]

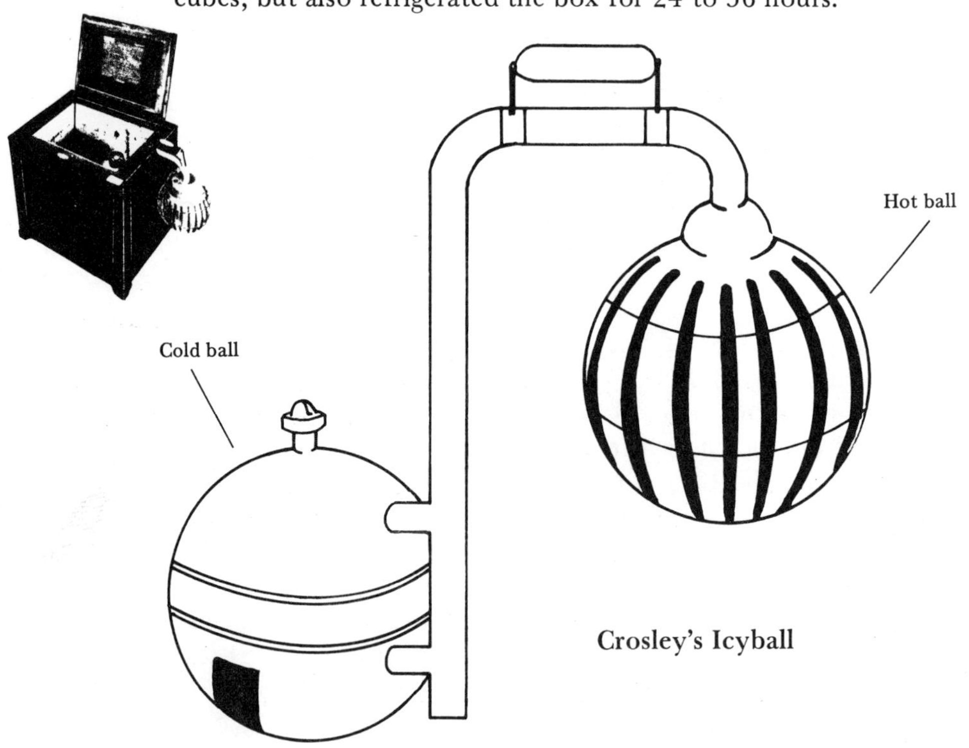

Crosley's Icyball

THE

CROSLEY

ICYBALL

The Crosley Corporation

"Scarcely had the iceman emerged as a regular visitor to the American home than efforts to eliminate him began. Among the first of these attempts was the Crosley Icyball Refrigerator made by the Crosley Radio Corporation of Cincinnati, Ohio. The unit consisted of a hot ball and a cold ball connected by a U shaped pipe through which ammonia vapors passed."

Air Conditioning and Refrigeration Industry Museum
Los Angeles, California.

If you are a skilled home craftsman, you may want to try your hand at building your own version of the Icyball. Stephen Peck, a marine surveyor in Clearwater, Florida, has built such a refrigerator successfully, and is offering to furnish a set of plans and a construction and operating instruction sheet for around $10. Write to him at P.O. Box 1592, Clearwater, Florida 33517.

INSTRUCTIONS
FOR OPERATION OF THE CROSLEY ICYBALL REFRIGERATOR

1. Remove ice tray from freezing tube in cold ball and place unit in draining position. The unit must be placed in the draining position before each heating. Draining will require no more than 3-5 minutes.

Before operating a new unit for the first time, drain and test the unit. Pour boiling water slowly on the bottom of the cold ball while the unit is in the draining position. Then rock the unit back and forth vigorously for several minutes.

To test for complete draining, pour more boiling water on the bottom of the cold ball. Place your hand on the bottom of the cold ball (which will be warm) and tip Icyball slowly from draining position to normal position. If the ball remains the same temperature, the unit is completely drained and ready for heating. If a change in temperature is noted, repeat the previous process.

It is also necessary to apply this procedure to a unit that has previously been heated too rapidly, or a unit that has not been used for some time.

2. When cold ball has completely drained, immediately submerge cold ball in tub of cool water, hooking catch over rim of tub and resting hot ball against wire bracket. Do not leave the unit in the drain position after draining has been completed. Be sure cold ball is completely covered by the water. Fill the reservoir of the kerosene stove. Light the stove and continue heating until the flame goes out. This will require about 90 minutes.

If you are using another gas, gasoline, or alcohol stove, adjust the stove so that the flame just touches the circular space inside the lower end of the fins of the hot ball.

Heat slowly so that after 90 minutes (and not before) a drop of water placed on the bend of the connecting tube nearest the cold ball will sizzle and boil immediately. Heating is complete when this sizzle test is obtained.

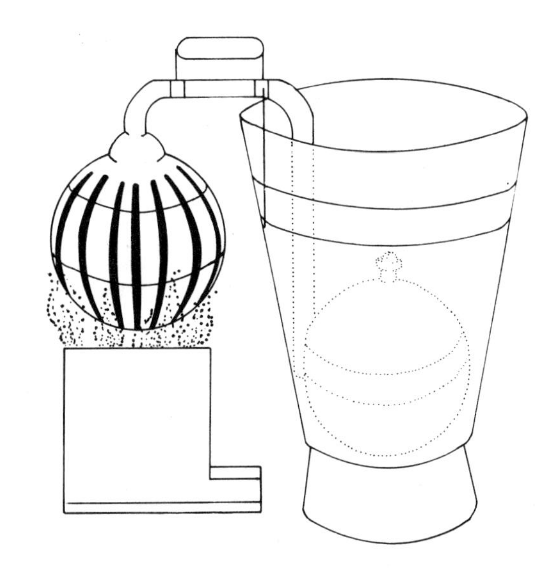

Heat properly.

3. When the heating is complete, remove the Icyball from the heating position, submerge hot ball in water, and hook catch over rim of tub with vertical tube resting against wire bracket. Leave unit in this position for five to ten minutes.

Cool the hot ball.

4. If the Icyball Stabilizer is used, remove cap and pour in 3 pints glycerine or any other odorless antifreeze, such as alcohol, salt brine or calcium chloride brine. Then fill with cool water and replace cap. Place stabilizer in bottom of cabinet so cold ball will fit in it when unit is in operating position. After the liquid in the stabilizer has once become chilled, it will keep the box cold and increase the length of the refrigeration period.

5. Place unit in cabinet, with hot ball outside and cold ball resting in bowl of stabilizer. End of vertical pipe will fit in stirrup inside cabinet. Fill ice tray with water or liquid to be frozen and slide tray into freezing tube. Remove tray by lifting up on handle until tray loosens, then draw it out.

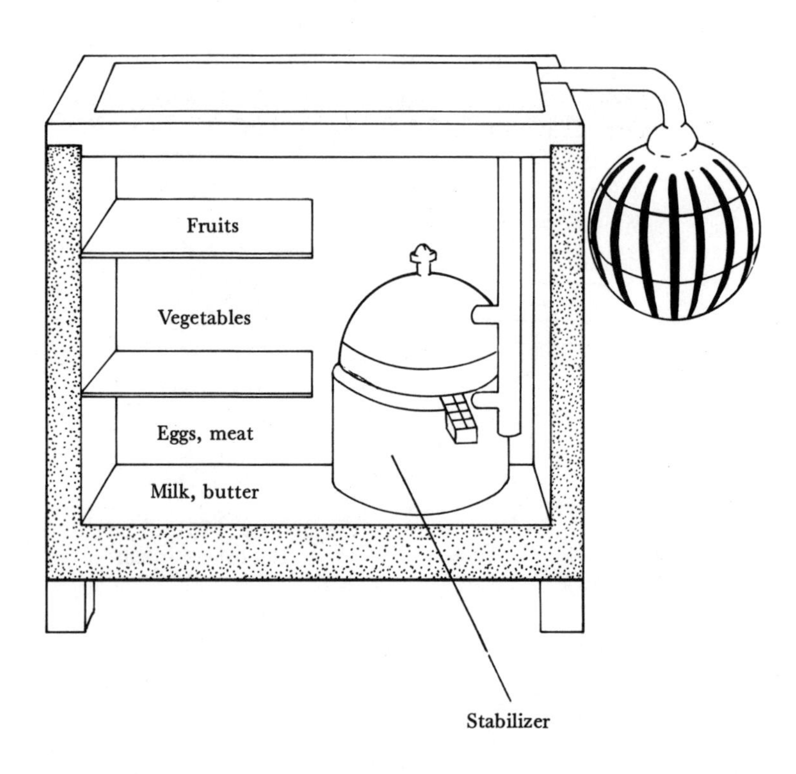

GENERAL ADVICE

When Icyball is put into service the first time or after a period of disuse, it should be heated at 12-hour intervals three times in order to cool the cabinet and stabilizer thoroughly. Once this is accomplished, reheat only when necessary (usually 24-36 hours), but more often in hot weather than in cold.

Best results will be obtained by heating the unit in the morning when the tub water is cool.

While the unit is cooling the cabinet, the hot ball will be warm because the heat from the cabinet and ice tray is being transferred to the hot ball and thence to the air.

•

The Icyball is very adaptable for solar use because the refrigeration mechanism is portable. It can be placed wherever solar radiation is the greatest for the heating period. And, since the heating period is relatively short, it would be possible to focus sunlight through a Fresnel lens, or bounce sunlight with a parabolic reflector, onto the bottom of the hot ball. Once heated, the unit could be placed in an *existing* icebox of the proper size, provided air circulation is sufficient around the generator end of the unit. On sunless days, the unit could be heated over the galley stove.

The Icyball was cheap, efficient and practical, and was popular until cheap electricty came into every neighborhood. The main disadvantage was that heat had to be applied at least once a day.

The Icyball suffered only one minor problem: some water vapor was carried over to the condenser/evaporator during the boiling out of the ammonia. Recent experiments have tried to find the best medium for solar cooling with ammonia. Ammonium nitrate, lithium nitrate, alkali halides, and sodium and lithium thiocyanate were some of the salts studied.

Although some of the nitrates had desirable qualities, they were capable of producing nitrogen and steam with *explosive vengeance.* These are, for obvious reasons, *not* the right salts for the home craftsman to employ. After much study, sodium thiocyanate was selected as the best salt to dissolve in ammonia for use in cooling and refrigeration. The solutions have suitable properties for this use, are chemically stable, inert, inexpensive, and can be used in iron vessels.

A laboratory test of the sodium thiocyanate-ammonia system was made using a 20-lb. blackened steel cylinder containing 12 lbs. of half sodium thiocyanate and half ammonia. An ordinary 4-ft. parabolic solar cooker was focused on the cylinder, which was connected by a flexible *pressure* hose to a smaller vessel immersed in water at room temperature. After heating for 4 hours, most of the ammonia had been driven into the small vessel. The generator was removed from its heat source and the small vessel was placed overnight in an insulated box containing water. When removed from its heat source, the sodium thiocyanate in the large cylinder cooled and re-absorbed the ammonia from the small vessel. The evaporation of the liquid ammonia cooled the contents of the insulated box and 9 lbs. of ice were produced.[6]

It is unfortunate that the Crosley Icyball is no longer in production because, at the present time, there is only one American company interested in the manufacture of an intermittent absorption-type refrigerator. This refrigerator, the "Fireice," a slightly updated version of the Icyball, is not now in production either. I was informed by the company spokesperson that only three or four of these units were ever manufactured. These prototypes were tested by the U.S. government in the jungles of Panama in 1973. The heat source of these prototypes was charcoal, and they reportedly performed well. However, the company never went into actual production, and they have no definite production plans at the present time (August 1978).

The only other company I have been able to locate that is interested in producing another version of the Icyball is The International Solar Power Co., Ltd., 26 Norrebakken, DK-2820 Gentofte, Denmark, but they have not yet started production.

There is a new type of solar refrigerator now being built by The Zeopower Co., 75 Middlesex Ave., Natick, Massachusetts 01760. It is much too heavy, large, unwieldy and expensive for most boats, but it incorporates a unique method for heat storage that is worth more study. The refrigerator's solar collector is filled with zeolite, an inexpensive mineral long used in water softeners. The zeolite has a tremendous capacity for heat storage. It works much the same way as the Icyball. At moderate temperatures, the zeolite absorbs the refrigerant vapor so completely that a semi-vacuum is formed causing the refrigerant to boil off, producing cold. After the zeolite is saturated, it is heated by the sun, driving off the refrigerant vapor, which condenses into a liquid and is ready to cool again.

The type of refrigeration, or lack of it, you choose will depend on your own personal needs and desires. But, each way we simplify our lives will, in turn, enhance them by increasing our strength and independence.

6

Solar Hot Water
How to Get it—How to Keep It

The simplest application of solar energy is heating water. Solar water heaters have been used for decades in many countries, particularly Israel, Japan and the United States. The use of solar energy to heat domestic water supplies has been common in this country in Florida and California since the 1930s. Solar water heaters were often the cheapest source of hot water before low-priced gas and electricity became available. Today, with the constantly rising cost of utilities, solar water heaters have come back into their own.[1]

While thousands of new solar water heaters are being installed daily, many of those original, circa-1930, solar water heaters are still in use today, and their far-sighted owners are the envy of their neighbors.

There are many solar water heating systems suitable for use on a small boat. All of them are simple, inexpensive and effective, so the one you choose will depend on how sophisticated a system you desire. If you are a purist who feels that anything more than a cold can of beans, a warm can of beer and a sleeping bag below decks is an affront, the bucket water heater is probably for you.

The Bucket Water Heater. If this is the system you choose, make sure your bucket heater is as efficient as you can possibly make it. Use a good quality galvanized metal bucket, with a sturdy handle. Wipe the bucket carefully with household white vinegar, then paint the bucket, inside and out, with flat black stove enamel. The vinegar treatment will prevent the paint from peeling off the galvanized surface.

Drill a hole near the lip of the bucket, smoothing the edges with a file. Then thread a 12-inch length of small diameter line through the hole and tie a knot in each end of the line. Fill the bucket with water and place it in the most convenient sunny location. When the water has reached the desired temperature, hang the bucket above your head (over the end of the boom or hoisted overhead on a halyard), pull the end of the line, and—presto!—a hot shower. Or pour the heated water into a thermos or insulated container for later use.

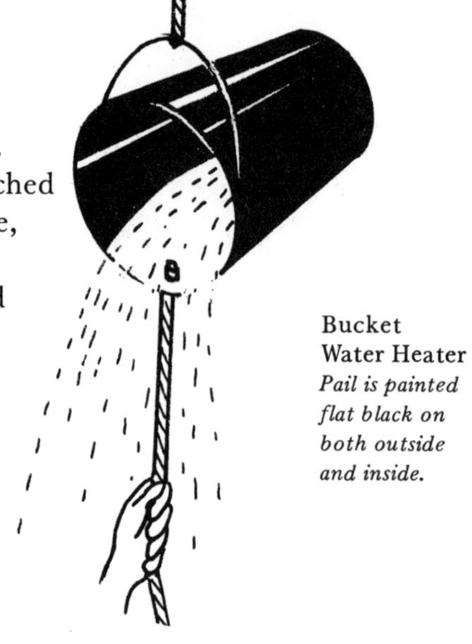

Bucket Water Heater
Pail is painted flat black on both outside and inside.

The Collapsible Plastic Jug. If you want a slightly more sophisticated shower, use a collapsible plastic jug with a shampoo hose and nozzle attached. Set the jug, filled with water (leave a little room for heat expansion), in a sunny location, on top of a piece of black-painted plywood, aluminum or cardboard. When the water is hot, the jug can be hung or gentle pressure can be applied, while you shower in luxury in the cockpit. Because of the fine spray from the shampoo nozzle, this method uses a minimum of water. You can get three or four showers – or more– from a 2½-gallon jug of water.

The Garden Hose Heater. Dockside hot water for the purist can be manufactured by merely coiling a piece of dark-colored garden hose in a sunny location. A good quality, 5/8 inch hose at least 100 feet long is best for this purpose. There should be a nozzle at one end of the hose and, once the hose is filled with water, the nozzle and the spigot should both be turned off. But remember, especially in the tropics, that heat expansion can burst a plastic hose on a hot summer day, so let a little water out to allow for this expansion.

This method may seem a little primitive, but Steven Smith of Vacaville, California, uses about 300 feet of plastic hose coiled on the roof of his home to produce hundreds of gallons of hot water daily for his swimming pool.[2]

A commercial version of *The Simple Solar Water Heater* shower can be purchased from many marine supply stores or through various discount catalogs. It comes complete with a heat sensor that changes color to let you know when the water is hot, and it holds approximately 2½ gallons of water. Your homemade version would cost only slightly less, but could be made in a variety of sizes.

Again, to hold the hot water for later use, place it in an insulated container. To retain the highest heat possible, place the entire jug in an insulated container.

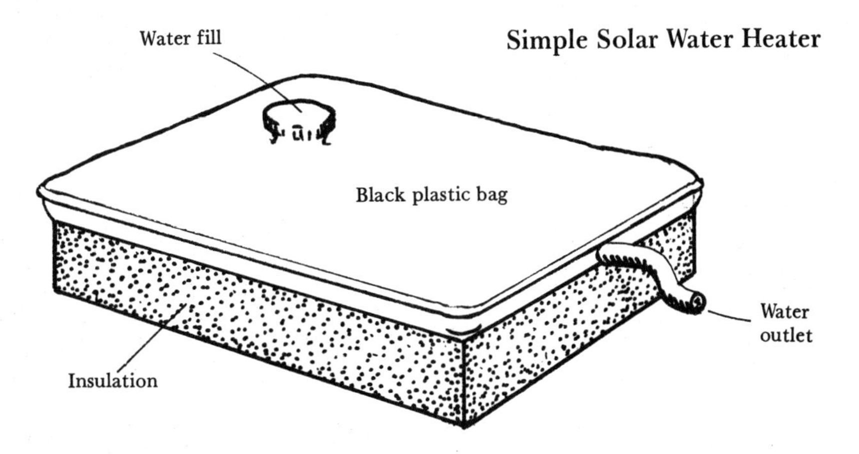

Garden Sprayer Water Heater. For the more sophisticated, a fine solar water heater can be made from a slightly modified common garden sprayer. These sprayers are readily available in several sizes and materials: 1½-gallon, 2-gallon, and 3 gallon capacities; in galvanized metal, stainless steel and polyethylene. These garden sprayers are equipped with a pump assembly with which to pressurize the tank, a bronze spray nozzle, which adjusts from a steady stream to a fine mist, and, usually, a pressure relief valve. The fine mist nozzle adjustment makes the most water-saving shower imaginable. My experience has been that a 3-gallon sprayer can offer ten *or more* showers, a couple of fresh-water rinses for all your diving equipment, and spot cleaning for a few gull messages, with enough left over to clean all the sunglasses on board.

The polyethylene sprayer has the advantage of not rusting in a salt-water environment, but plastic is a poor conductor of heat; therefore, a metal sprayer is a better choice for use as a solar water heater. Stainless steel is best, of course, because of its rust-resistant quality, but a good quality galvanized sprayer can be very serviceable and is less costly.

The garden sprayer has been used for years as a simple shower by cruising sailors. Our job now is to adapt it for the most efficient possible *solar* use. First, paint your sprayer with flat black, heat-resisting stove enamel. Next remove the hose, which is usually white and attached with inferior clamps that will rust your first day at sea. Replace the white hose with black pressure hose and attach it with small stainless steel clamps. The hose and clamps can be purchased from any auto supply store. Next, place your sprayer in a sunny location. It can be secured to the lifelines or shrouds, tied to the mast or placed in any location that is convenient for you. When you are ready to use the hot water, tighten the cap securely, pressurize the tank by pumping, adjust the nozzle to the spray you desire, and enjoy a terrific shower.

Since the tank is lightweight and portable, it can be used anywhere after the water has been heated. A self-bailing cockpit is always a good place for a shower, or the head area can be fitted with a shelf or strap to accommodate the preheated tank for a shower in real privacy. Aboard *Tevake,* a sprayer is lashed to the lifelines just outside a porthole that opens into the head area. By pulling the sprayer hose through the open porthole, we can shower in privacy without even moving the sprayer. A garden sprayer is also a handy addition to the galley for a hot-water rinse for dishes washed in cold salt water.

But again, remember that the water won't stay hot once the sun goes down, without a little help from you. The heated tank should be placed in an insulated container of some kind. One of the insulated bags designed to keep food hot or cold, or a well-insulated ice chest, could be used, or the sprayer could be wrapped in a blanket or heavy towels,

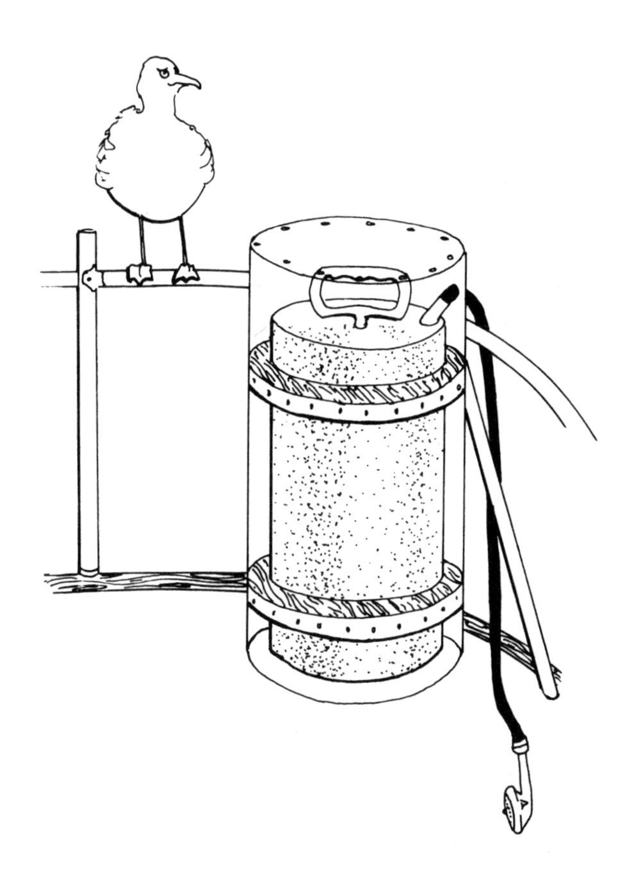

Garden Sprayer Water Heater

or covered with a space blanket. Experiment a little and find out what works best for you.

The greatest disadvantage to the simpler type solar water heaters is that they have no provision for preventing convective heat loss. Convective heat loss may be no problem on a hot summer day, but on a sunny winter day or when the wind is blowing, your solar water heater could lose as much heat as it gains by convection back to the surrounding air. This can be overcome very simply by making a cover for the heater from clear heavyweight polyvinyl film. This film can be purchased from any large garden supply store. It can be sewed by hand or machine, or tacked onto a simple frame. When sewing plastic film, use dacron thread and an upholstery or sail needle. If sewing on a machine, run a narrow bead of machine oil along the top layer of the plastic film, so that it will slide easily under the pressure foot.

Whether you try the bucket, the collapsible plastic jug, or one of the other simple solar water heaters, you will be convinced of the power of the sun, and may want to install a more permanent type solar water heater. If so, it is a fairly simple matter to design and build one yourself, in a size and location that exactly suits your needs.

Experts seem to agree that one-third of the water we use each day is *hot* water, but there are widely varying estimates of daily water consumption for the "average" person. In the U.S. the estimate is 28 gallons per day in public housing and 41 gallons per day in luxury apartments. In Australia, the daily average is 10 gallons per person.[3] Shipboard use of water is much more limited than any of the above estimates. Exactly how much is used depends on how limited the supply is – but probably averages 2 to 5 gallons per day per person. Since only one-third of that amount needs to be hot water, it is apparent that most small boats can accommodate a solar water heater of adequate size.

The simplest water heater is a horizontal black tray of water set in the sunshine. But, as soon as the tray of water heats up, it begins to lose heat by evaporation, by radiation and convection to the surrounding air, and by conduction to the materials from which the tray is made. Therefore, solar water heaters are fitted with transparent covers to trap the heat, and they are placed in an insulated box to prevent *conductive* heat loss.

Since the heaviest demand for hot water is early in the morning and late in the evening, the water that is heated when the sun is shining must be stored. This can be accomplished in several ways. The simpler the system, the more user participation is required. The more complex the system, the less

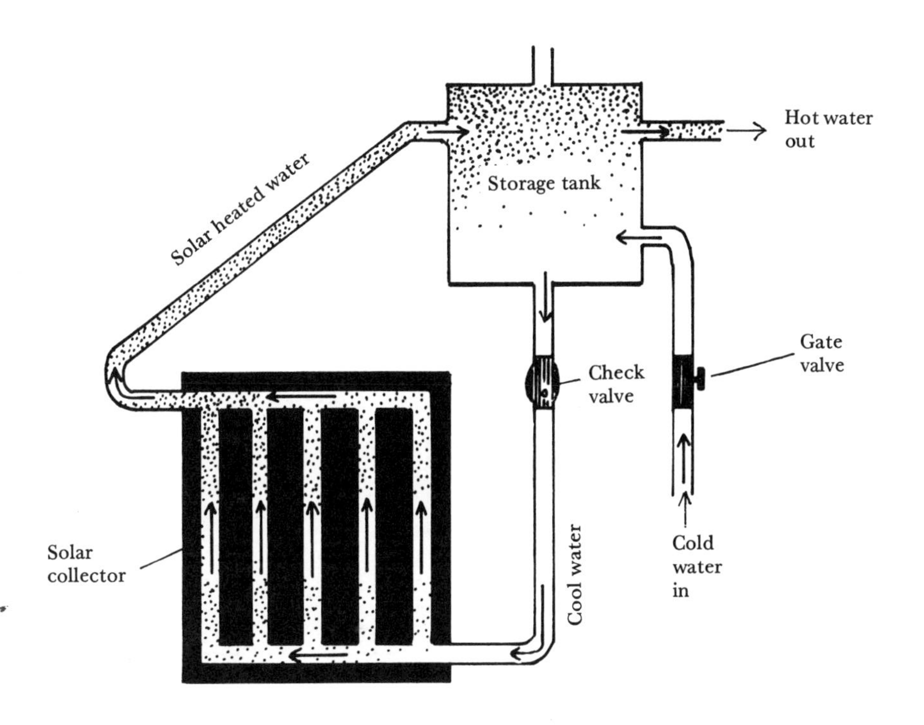

Remote Storage Thermosiphoning Water Heater

active participation is necessary, but the more expensive the system is.

A separate, well-insulated storage tank can be used to store the water heated in a collector and, if it is located *above* the collector, the hot water will be transferred from the collector to the storage tank by natural thermal circulation. The water rises as it heats, while the cool water settles to the bottom. As long as the sun is shining, water will continue to circulate. Heated water from the collector will flow through a hose to the top of the storage tank as the cooler water at the bottom of the storage tank flows back into the collector to be heated and to rise gradually to the top of the tank. If the storage tank is well-insulated, the water will stay warm overnight.

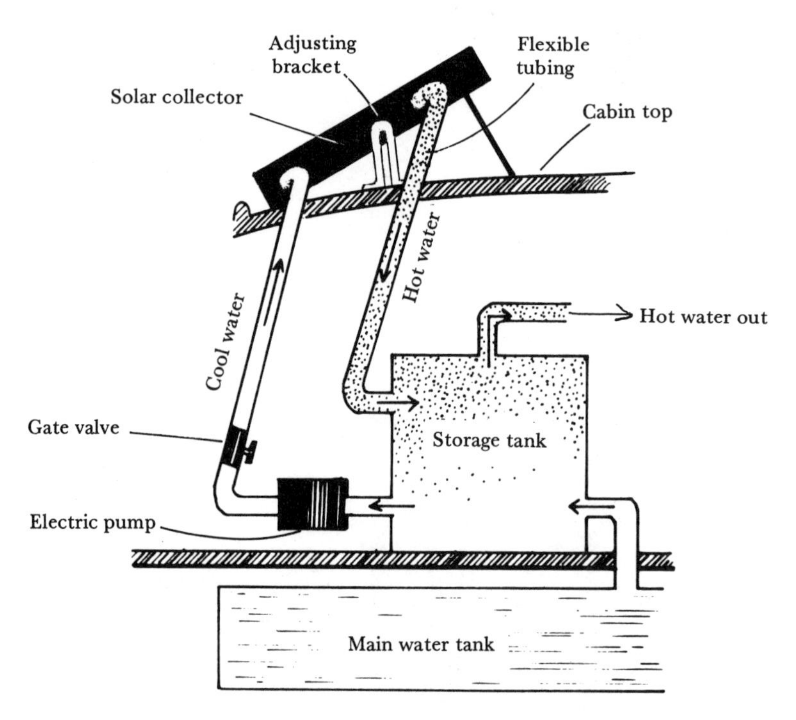

**Remote Storage
Water Heater With Pump**

If the storage tank cannot be placed above the collector, natural thermosiphoning will not take place; therefore, a pump must be used to transfer the hot water. See the comparative illustrations on page 75 and above. Both of these systems are worthy of study when refitting an existing boat, because they can be incorporated satisfactorily into an existing pressure water system.

The only difference between the solar water heater shown on page 75 and the one illustrated above is the location of the storage tank. Since no electric pump is required for the thermosiphoning water heater, it is the more energy efficient of the two systems.

Exact instructions for building a thermosiphoning solar water heater are included in Chapter 11.

For the best possible performance from any water heater, it must be heavily insulated. Not only the back and sides of the collector and the entire storage tank should be insulated, but also the attached tubing.

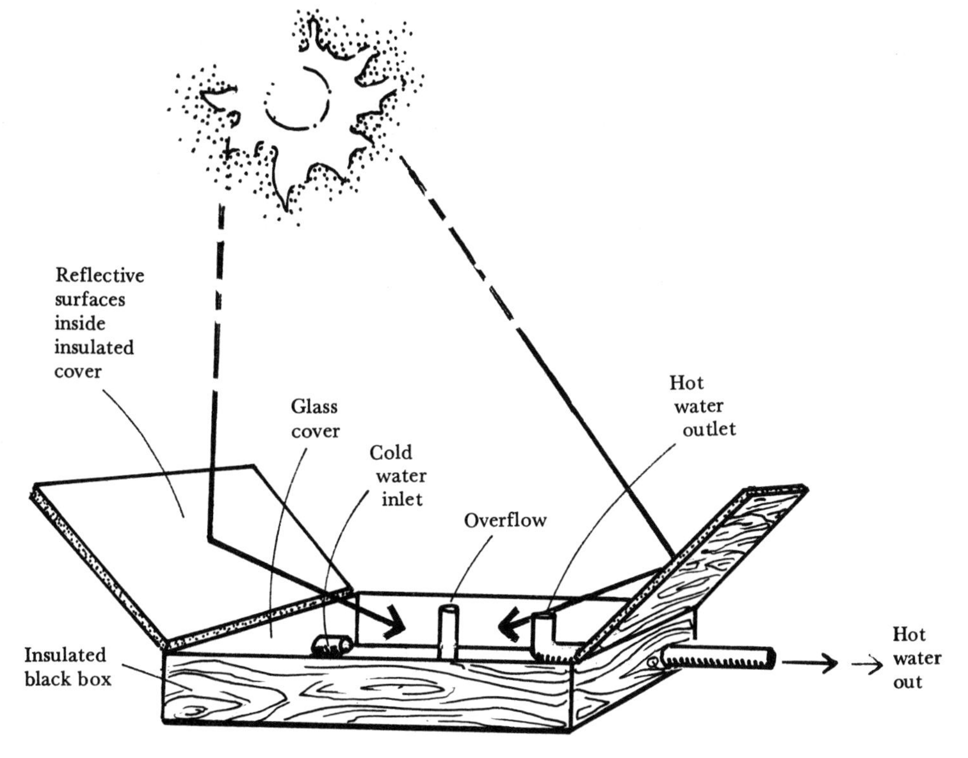

**Storage Type Water Heater
With Reflective Cover**

In cooler climates, locating the collector horizontally is less efficient because the sun is lower in the sky, expecially in the winter. To overcome this problem (if the collector is fairly small), it can be mounted with an adjustable bracket as shown in the illustration on page 76. The collector can then be set at the optimum angle. Aluminized reflectors could also be added to reflect additional sunlight onto the collector.

Where space is limited, it is best to use a storage-type heater, or one that functions both as a collector of solar energy *and* a storage unit for the heated water. The greatest problem with these storage-type heaters is their excessive heat loss at night and on cloudy days. The best remedy for this is to insulate the back and sides of the unit heavily, and to use a removable heavily insulated lid to put over the glass cover when the sun is not shining. The insulated lid could serve another purpose: if it were lined with a reflective film, it would gather additional sunlight and bounce it onto the surface of the heater when open. The only disadvantage to this system is that it requires some participation by the user, to open the insulated cover.

The illustration shows another type of thermosiphoning water heater developed and being used in the West Indies. The collector is a blackened glass-covered metal container. On one end, this collector bulges out to form its own storage tank. A baffle in the center of the container assists the thermosiphoning action. There is extra insulation in the storage portion in order to keep the water warm overnight.[4]

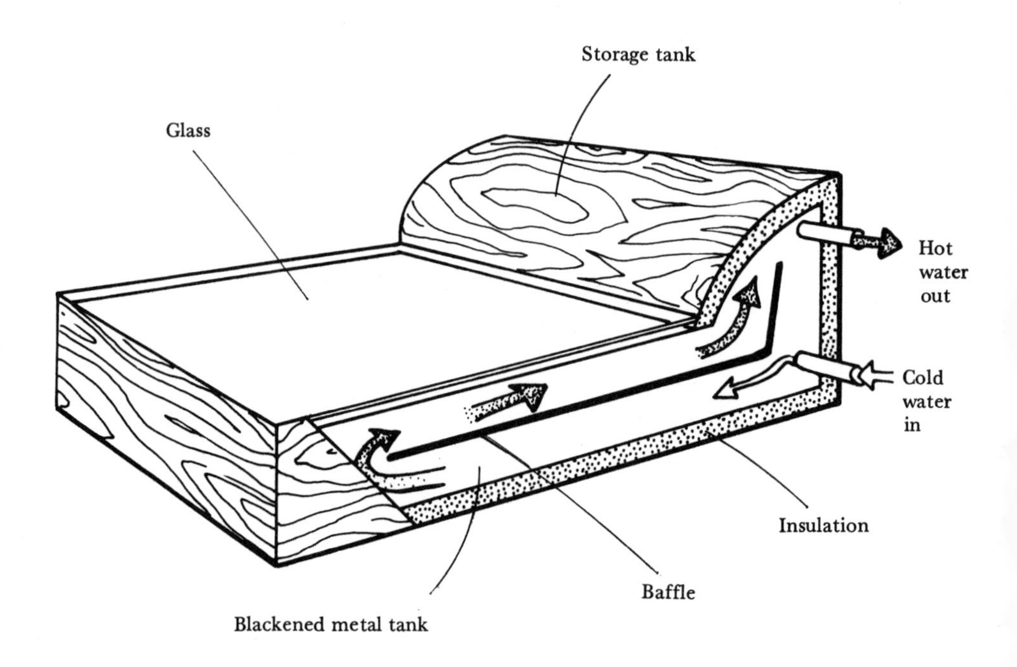

Storage Type Thermosiphoning Water Heater

If your solar water heater is designed for use where freezing temperatures are even a possibility, safeguards must be included to prevent the water in the solar heater from freezing. When the air temperature drops below 32 degrees F, freezing water can burst pipes or collector channels.

Your thermosiphoning collector can be protected by:

Heating the absorber.
Covering the collector with movable insulation.
Draining the collector.
Using antifreeze.

There is an inexpensive product called "heat tape," available through stores similar to Sears Roebuck or Montgomery Ward, which can be fastened to the back of the absorber plate. This tape looks like an extension cord and, with the help of a thermostat, causes an electrical current to heat the absorber plate every time the temperature falls below 35 degrees F. This tape is a good *dockside* solution in areas where freezing seldom occurs, but in more severe climates, the operating cost would be prohibitive.

Movable insulation is a cheap, effective solution to the problem of freezing, but requires user participation.

Draining is effective, provided the system has been designed to drain completely. Again, this requires user participation.

A SOLAR WATER HEATER USING ANTIFREEZE

The most effective protection against freezing without any extra user participation is antifreeze. However, this complicates the design and increases the cost of the system. The antifreeze must be isolated from the water; therefore, a heat exchanger of some kind must be used to transfer the solar heat from the antifreeze to the hot water supply. A heat exchanger could be made from a closed coil of copper tubing immersed in the storage tank, or a smaller tank inside the larger storage tank. Copper tubing containing the antifreeze could even be wrapped around the outside of the storage tank. Be sure to provide a vent to avoid overpressure in the

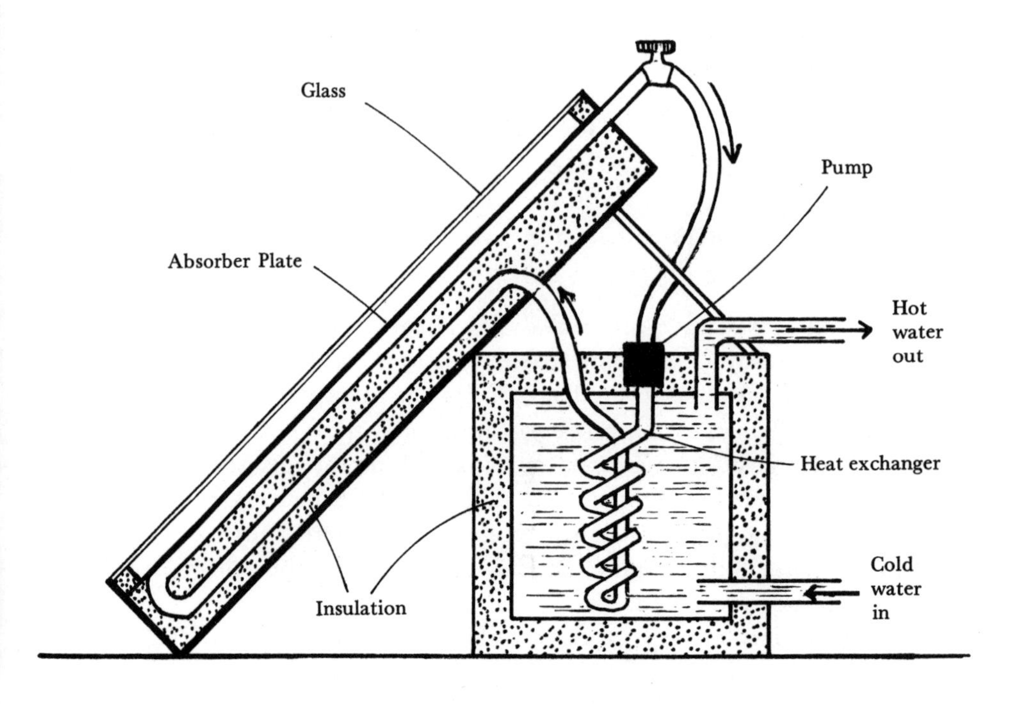

**Solar Water Heater
Using Heat Exchanger Containing Antifreeze**

heat exchanger. The most important point is to be sure antifreeze never enters the water supply.[5]

In climates where freezing weather is common during the winter months, auxiliary heating will be required for your solar heater, as well as the use of antifreeze. In this type of system, water is solar-heated with a heat exchanger, and then passed through a thermostatically-controlled conventional water heater where the preheated water is brought up to the appropriate temperature. The preheating by solar energy will greatly reduce the amount of time the conventional water heater requires to heat the water, thereby greatly reducing the amount of electricity required when used dockside.

Regardless of your climate, if you already have a conventional water heater incorporated in a pressure system aboard your boat, consider it a backup on cloudy days to your solar water heater which pipes heated water through the conventional heater before it passes through tubing to the hot water faucets.

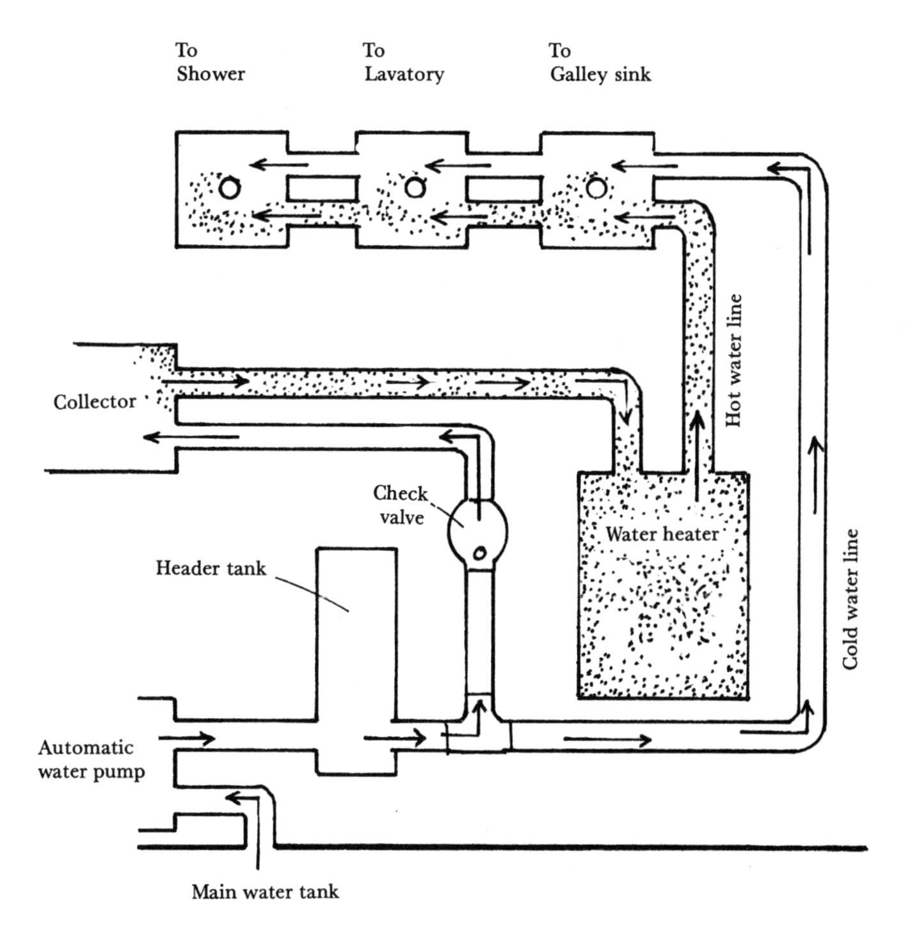

Auxiliary Water Heating System

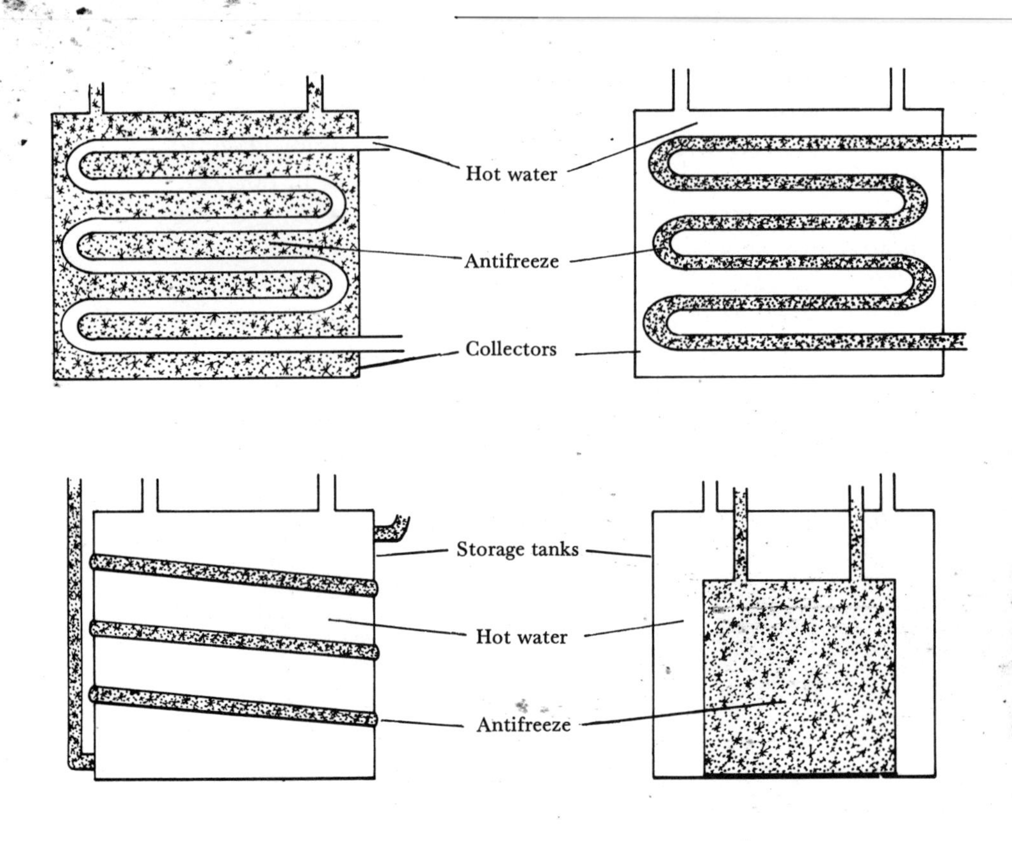

Heat Exchangers for Solar Water Heaters

Various types of simple heat exchangers are shown above.[6] The heat transfer fluid used is usually ethylene glycol or any other readily available antifreeze solution.

Building a solar water heater is a perfect place to put recycled materials to work. Storage tanks can be made from any large discarded galvanized can (provided it can be thoroughly cleaned), or from the entrails of a discarded conventional water heater. A new heat exchanger could be made from salvaged air conditioner or refrigerator coils. Salvaged scrap lumber, panes of glass, copper tubing, newspapers, styrofoam, insulation out of discarded stoves or refrigerators, can all be put to use to reduce the cost of your new solar water heater. There have been a number of articles in past issues of *Mother Earth News* that describe how to build various types of solar water heaters using discarded materials.[7] Although most of these water heaters are too large for use on a small boat, they do show some unique ways of keeping costs down with recycled materials.

Instructions for building two different types of 5-gallon solar water heaters are included in Chapter 11: a thermosiphoning heater with a remote storage tank and a storage-type collector. Before building your solar water heater, however, first determine whether you will encounter enough freezing weather to warrant the use of antifreeze and a heat exchanger. If so, use the examples shown on pages 80 and 82 as a guide. In warmer climates, you will be able to use a simple gravity feed storage-type collector, with a provision for covering or draining in case of an occasional freeze.

Next, you must study your available space to determine whether you have an appropriate location for a thermosiphoning-type heater. If space is very limited and a remote storage tank is necessary (it requires the use of a pump), consider incorporating the preheated water from your collector or storage-type collector into your existing pressure water system, using your conventional water heater as the storage tank.

You will enjoy your solar water heater most if you keep the system as simple as possible, consistent with your local weather conditions.

WARNING: With any of the water heaters shown in this chapter, it *is possible* to receive a nasty burn on a bright summer day when solar radiation is at its peak. So *don't* use your solar-heated water without first testing its temperature.

7
Put the Heat of the Sun to Work on Your Boat

Cave men did it. Pueblo Indians did it. Why shouldn't we do it, too? Using the heat of the sun is only a matter of living more in tune with the environment.

There are many examples in nature of simple, direct methods to compensate for temperature variations of solar radiation throughout the year. Some flowers open and close with the rising and setting of the sun. Many animals find natural shelter from the heat of summer, and bury themselves in the earth to stay warm in the winter. Primitive man, more in touch with nature than modern man, learned by observation how to tune in to his environment. Caves, for instance, have nearly constant temperatures and humidities year-round; so do the adobe buildings of the Pueblo Indians. If we will just allow our intuitive understanding and appreciation of nature to take over, we will soon have the sun working for us.

The solar boat should be able to moderate both daily and seasonal extremes of temperature. On sunny winter days, it should be possible to open the boat to receive the sun's heat. At night, it should be possible to close out the cold and keep in the heat that has been absorbed by the boat during the day. In the summer, the opposite should be true. It should be possible to close off the sun during the day and open the boat at night to release any heat that has been absorbed, back into the night air.

There are three basic requirements for heating your boat directly from the sun: the *boat itself* must be a solar collector; it *must* be able to *store* heat; it *must give up its heat reluctantly*.

The boat, like any other good solar collector, must absorb a large part of the solar energy it receives. This amount can be maximized by orientation, color, generous placement of portlights and the use of shading.

In the cooler climates, south sides of fixed structures receive almost twice as much solar radiation in winter as in summer. The east and west sides of these same structures, however, receive two and one-half times more solar radiation in summer than in winter. In the warmer climates (latitudes less than 35 degrees), fixed structures gain even more on their south sides in winter than summer, and east and west walls can gain two or three times more heat in summer than south walls. For these reasons, proper orientation could make it much easier for you either to use or repel the sun's heat. If you live aboard your boat at a dock much of the time, the best orientation to the sun, as discussed at length by Victor Olgyay in *Design With Climate,* is shown in the diagram for northern latitudes. Southern latitudes would require the reverse orientation.[1]

Your boat is not a fixed structure except when at the dock, but, being a good navigator, you will always be aware of her exact position in relation to the sun, so you will be able to plan in advance to compensate for a less than ideal orientation to the sun.

Most texts on solar energy contain page after page of tables from which to make energy computations. From these computations, it is possible to determine almost exactly how much energy is available, how much is being received, how much is absorbed and emitted, how much can be used and how much will be lost — for fixed structures in every geographical location. Since boats are seldom fixed structures, we must take a more practical approach to the problems of solar design by using, wherever possible, portable systems that can be applied when the boat is in *any* position and by generous use of both access and resistance from all directions.

Ideal Orientation For Four Different Climates

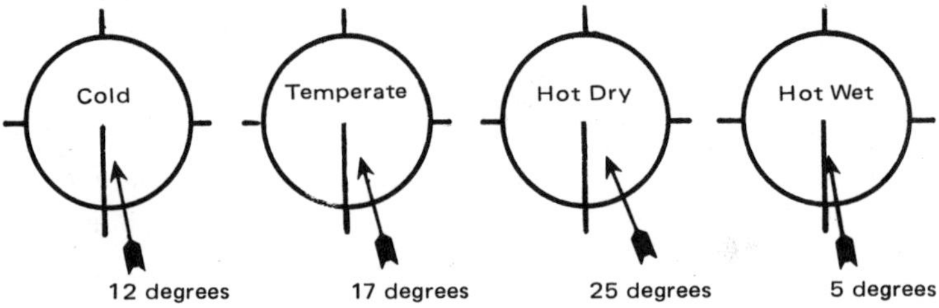

Cold — 12 degrees
Temperate — 17 degrees
Hot Dry — 25 degrees
Hot Wet — 5 degrees

Most of the tables in this book contain information applicable primarily to boats and are extracted from more complex engineering tables. If you want further information, or if you prefer to use the engineering tables directly, their sources are listed in the bibliography.

The *color* of the decks, topsides and hull of your boat greatly affects how much heat will penetrate the boat. Anyone who has stepped barefoot on a teak deck at midday can testify that dark colors absorb more heat than light colors. It would be ideal if we could paint the boat bright white for summer to reflect the sun's rays, and flat black for winter to absorb the sun's heat. But, since that is not practical, color, too, will be a compromise. Teak or colored decks and black or dark-colored hulls will be best for the cooler latitudes. White or light hulls and decks will be best for the latitudes closer to the equator. If your boat is a constantly cruising, ocean-going yacht, moving through many latitudes and many seasons, you will have to give considerable thought to an agreeable compromise in color.

Although the colors used on the boat and its orientation to the sun are significant, the size and placement of windows or portlights are even more important. These openings allow for the passage of the sunlight and air necessary to properly regulate temperatures inside the boat. Glass in these openings, besides reducing the amount of electricity needed for lighting, captures heat through the "greenhouse effect." Glass and the hard, clear plastics readily transmit visible (short-wave) radiation while holding in the resulting thermal (long-wave) radiation. This produces an effective heat trap, which you can use to warm your boat.

South-facing portlights will greatly improve the quality of the boat as a solar collector because they will receive more than twice the solar energy in the winter when it is needed as in the summer when it is not desirable. This is true because the sun is lower on the horizon in the winter, is closer to the earth, and the sunlight strikes the portlights more nearly at right angles. Effective shading for *all* portlights in the summer will increase their effectiveness.

Greenhouse Effect Through Boat Windshield

The type of glass (clear or reflective, single or double pane) used in portlights will also have a significant effect on heat gain. In the past, glass was the only transparent material used in solar collectors. But plastics are available now that are very close to glass in their ability to transmit light while holding heat. In addition, they are lightweight and will not shatter. So, you may use glass interchangeably with any type of acrylic plastic glazing. However, keep in mind that plastic glazing eventually clouds from repeated exposure to the sun and has to be replaced. Glass, on the other hand, sometimes breaks and has to be replaced.[2]

A *single* pane of *clear* glass has a 97% heat gain in the summer and a 68% heat gain in the winter. While a *single* pane of *reflective* glass reduces the heat gain to 58% in the summer, heat gain becomes only 19% in winter. *Double clear* glass gains 83% in the summer and continues to gain 68% in the winter. *Double* glass, using *clear* glass outside *and reflective* glass inside, gains 50% in the summer and 42% in the winter, which could be a happy combination, depending on your geographical location.[3]

In many climates, keeping the summer sun out is imperative; the sun can be a killer as well as your helpmate. Reflective glass definitely does reduce solar heat gain and there are many brands of reflective film on the market that can be applied to existing windows and portlights. But this may prove to be more of a disadvantage in winter than it is an advantage in the summer, except under certain circumstances. If large glass areas are involved, such as on houseboats or cruisers, summer heat gain could be a significant problem. Also, if your boat winters in dry dock or under canvas somewhere, the ability of windows or portlights to gain heat in the winter is not important. In most cases, however, double clear glass and portable shading devices offer the best year-round solution to proper heat gain through windows and portlights.

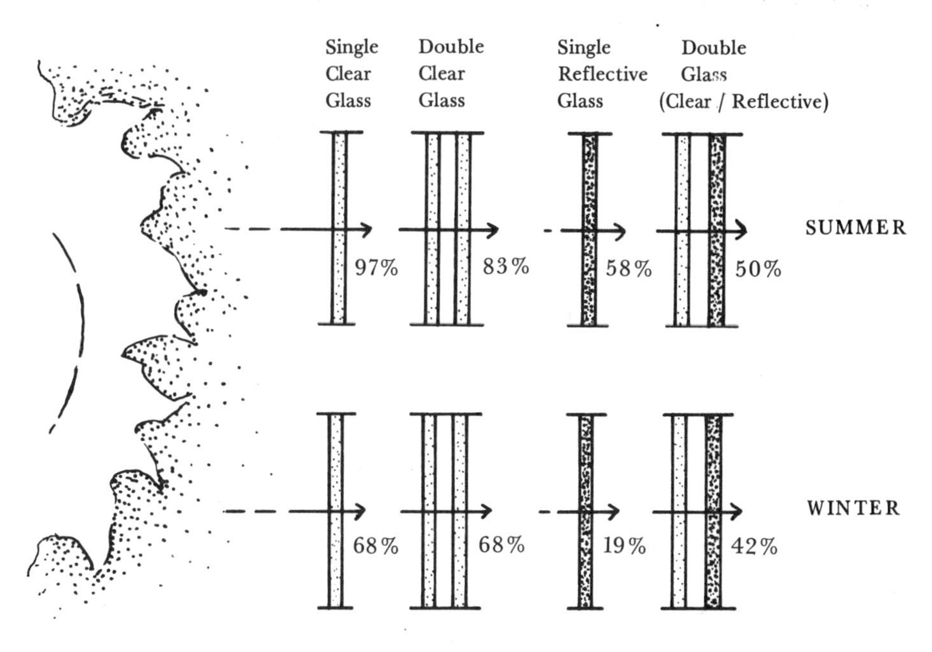

Seasonal Heat Gain Through Different Types of Glass

Shading can be accomplished in many ways. Small, permanently-mounted, plastic or metal shields or eyebrows are manufactured to fit different sizes of portlights. These shields protect the portlights from rain and spray and perform the function of a fixed awning. Other fixed metal or canvas awnings could be used over areas of glass, but they have certain disadvantages. If the awnings are designed for optimal shading in mid-August, they will cast the same shadow in May when an increased heat gain would have been welcomed. Therefore, movable shading devices are preferable. Dodgers and bimini tops that fold down and can be equipped with side curtains, large harbor awnings or boom tents, or permanently-mounted, folding canvas awnings will be more serviceable in providing shade at the right place and at the right time. Exterior shutters, roll-down blinds and interior thermal curtains also can help prevent heat gain from solar radiation through hull openings.

The Boat as a Storehouse of Heat. A vital part of a solar boat is its ability to soak up and store heat. How well the materials from which the boat is built hold heat is the key factor. Solar energy penetrates the hull, deck, topsides, and portlights of the boat, and the solar heat is absorbed by the air and its surrounding materials inside the boat. The air inside the boat warms first and distributes that heat to the materials that contain it. Once the surrounding materials have absorbed all the heat they can, the air continues to heat, eventually making the interior of the boat uncomfortable. Therefore, the larger the heat storage capacity of the materials from which the boat is constructed, the longer it will take for the air to reach an uncomfortable temperature inside the boat.

If it is cold when the sun goes down, the boat begins to lose heat to the cool night air, and if it has not stored enough solar heat during the day, auxiliary heating devices must be used to keep the crew and passengers warm. If the material from which the boat is constructed has a large enough capacity for heat storage, however, that warm material will replace the lost heat as the inside air cools and the boat will maintain a more or less constant temperature. However, the heavy materials that store heat best are poor insulators, and the insulating materials are the ones that prevent that stored heat from escaping. So in order to have the most effective thermal mass, it is best to insulate both the inside and the outside of the boat.

The same boat that has enough thermal mass to store heat can also store coolness at night for use during a hot summer day. At night, when outside air is cooler than it is during the day, ventilation will cool the air inside the boat and its surrounding materials. Since they will be cool at the start of the following day, they will absorb and store more heat before they, too, become warm, thereby cooling the inside air as they absorb heat from it. With enough thermal mass, it will take a long time before the air inside the boat reaches uncomfortable temperatures.

To get a general idea of the thermal mass of your boat, refer to Table 3, the "Heat Storage Capacity Table." A ferrocement boat usually has great thermal mass, as does the wooden boat, the steel-hulled boat (especially if the hull has been finished with troweling plaster), and the aluminum boat. Fiberglass boats have less thermal mass as a general rule; however, the hull material is only one factor involved. The materials used for finishing both the inside and the outside of the boat also contribute to the thermal mass of the boat. Your own observations are probably more accurate than any tables for determining the ability of your particular boat to hold heat. Who would know better than you if the boat is hot as blazes in the mid-summer sun and equally cold on a winter's night?

If your boat responds slowly to outside temperature variations, it will need little in the way of auxiliary heating or cooling devices. On the other hand, if your boat tends to become uncomfortably hot or cold, you should explore ways of either adding or enclosing existing thermal mass in a sandwich of insulation.

The Boat as a Heat Trap. We know how best to collect heat and store it; now we must reduce the escape of that heat.

The escape of heat from the boat in winter is considered heat loss, while the absorption of heat into the boat in summer will be called heat gain. Our job will be to retard this movement of heat into and out of the boat in order to keep the temperature comfortably constant. Our job is made easier by the fact that anything we do to reduce winter heat loss will also reduce summer heat gain.

Heat escapes from a boat in three ways: by convection of air through openings in the exterior surface; by conduction through the cabin sides, coach roofs and cabin soles; and by conduction through closed hatches and portlights. The most rewarding thing you can do toward solarizing your boat is to attack all three types of heat loss. The result will be a decreased demand for or even total elimination of auxiliary heating or cooling devices.

If your boat leaks water during heavy rains or heavy seas, then you can be sure it also leaks heat in those same places. This is the time to caulk, rebed stanchions, weatherstrip, and see that all ventilators or air scoops have movable or removable air-tight covers. Ports and windows are especially vulnerable to air and water leakage. If necessary, they should be removed and rebedded, replacing the rubber weather seal in the process. Extra care should also be taken in weatherstripping hatch openings and exterior cabin doors.

The only way to reduce the conductive heat loss through the cabin sides, coach roofs and cabin soles is by adding resistance to the heat flow with insulation. Since warm air rises, you should add the most insulation to the roof, then the cabin sides, and finally the cabin soles. And, wherever possible, insulation should be added to the exterior surfaces of the boat. The type of insulation you use will depend on many variables: the contours of the surface; how much space is available; how much weight you are willing to add, etc. Fortunately, the better insulating materials are very light in weight. To get an idea of the comparable qualities of various insulating materials, refer to Table 2, "Resistance of Common Insulation Materials." You will notice that air spaces are good insulators, as are materials that have small air spaces trapped inside.

Whenever possible, if excessive weight is not a problem, the exterior surfaces of the boat should be insulated. This could be done by the addition of a laminated teak or canvased plywood deck, cabin top or wood-laminated hatch covers. Because of its color and density, teak is more suitable for colder latitudes, but could be a welcome addition even in subtropical climates if it is kept treated or oiled in the winter and allowed to "go light" during the summer. If you choose plywood, be sure it is the best quality "marine grade" plywood, and that all joints are well-sealed. However, laying a plywood deck is a rather complex task and not really a do-it-yourself project. The ease with which teak can be applied, and its superior resistance to rot, may compensate for its high cost.

Insulating the interior of your boat will be a much simpler task. Acoustical tiles, corks or corkboard, dropped ceilings, carpeting, all conventional interior insulators, can be used where exposure to wind and water is not a problem. Lockers and cabinets with their accompanying air spaces also add insulation. An asphalt or latex compound similar to undercoating can be brushed onto fiberglass hull interiors for insulation; however, a fiberglass hull that is constructed with a core of balsa, plywood, honeycomb or foam will need little added insulation.[4]

A boat is further insulated by the water in which it floats. Temperatures in the keel area of the boat will remain almost constant and near the temperature of the water. If the water is extremely cold in your area, this too must be taken into consideration and adequate insulation provided below the cabin sole.

Reducing conductive heat loss through closed ports, windows and hatches requires insulation, too, but used in a different way. The heat loss through a single pane of glass is 30 times that of a well-insulated cabin wall. Just using double glass with a small air space between the panes will cut the heat loss almost in half. If a shutter is then added to the double glass, the heat loss is cut to about one-tenth the loss experienced with single glass. If you don't have room for shutters, heavy insulating cloths can be used, but they must be held tightly against the glass. Heavy drapes used without insulating cloths are poor insulators because they don't prevent the movement of cold air through the folds. However, heavy drapes used *over* tightly-fitted insulating cloths are very effective insulators.

Wherever possible, hatch covers should be insulated, too. If this is not possible, heavy insulating cloths again can be snapped on or held in place with velcro. The worst culprit in heat loss is the wind. When the door to the main hatch is opened for entry or exit, vast amounts of heat escape into the air. The only way to prevent this is to provide some sort of wind protection around this opening. A completely covered cockpit enclosure is the best protection; a large awning or a dodger with side curtains that shield as much of the opening as possible would also be helpful.

If you are a cruising sailor, or if you live or entertain aboard, the modifications suggested in this chapter can add immeasurably to your comfort level. The small amount of weight added won't seem important when compared to the enormous advantages of owning a "solarized" boat.

The design principles for using your boat as a solar collector, solar storehouse, and a solar heat trap have been explained in this chapter. Exactly how to put these principles to work is discussed in the following chapter.

8

The Nitty Gritty of Direct Solar Heating and Cooling

When preparing to receive the sun, the best place to begin is with a dry boat. A boat that leaks water also leaks heat. So let's seal the boat first to prevent this from happening, as well as to provide a dry area with which to work.

It is sometimes difficult to determine exactly where a trickle of water has originated. There are products on the market that can be used for tracing a leak; they leave a telltale dye marker. Or, you can mix a little household bluing in a pail of water and pour it over one small suspected area at a time until all leaks have been found and stopped. Normal working stresses make it necessary to rebed stanchions, chainplates and other deck hardware periodically. This is a good time to perform this task as preventive maintenance.

It takes two people to remove and replace through-bolted deck hardware: one person above deck and one below. It is easiest if the person on deck turns the bolts out while the person in the impossible position below deck holds a wrench on the corresponding nut to prevent its turning. Once the hardware is removed, both the hardware and the deck must be scraped and thoroughly cleaned of any old bedding compound with a suitable solvent. Clean the old nuts, bolts, washers and backing plates as well, and replace any that are badly rusted or bent. Rust can be considered a contagious disease. Apply a generous amount of bedding compound or caulking to the holes in the deck, replacing the deck hardware without disturbing the compound any more than necessary. Replace the bolts and hold them in place while the person below deck pushes on the backing plate and washers. In order to get a *fool-proof seal,* it is important that the person above deck hold the bolts in place with a screwdriver to prevent their turning while the contortionist below deck turns the nuts on. As the nuts tighten against the washers, compound should ooze out on all sides, above and below deck. If it doesn't, you haven't used enough compound. Take the hardware off and start again, using more compound this time. Don't be too fastidious about this operation; it's one you won't want to repeat next year. Once the nuts are all tightened, clean away the excess compound with the recommended solvent.

Windows and portlights should be rebedded in the same manner if they have shown any signs of leakage around their perimeters. Inside sealing gaskets should be replaced on opening windows if they leak through their seal.

Fixed windows should be replaced with opening windows wherever possible. The more available openings you have into your boat, the more flexibility you will have in heating and cooling without excessive auxiliary help. This will help compensate for a less than optimum orientation to the sun.

A second pane of glass or acrylic plastic should be added to windows and portlights. These second panes can be cut to size, pushed into place with an air space between the panes, then sealed with foam weatherstripping around the perimeter of the window. These storm windows can be permanent or temporary, depending on your geographical location.

To further hinder the escape of heat through windows and ports, detachable exterior shutters can be used. These shutters can be made of plywood, sheet polyurethane foam, or any available stiff insulating material. These shutters will also be useful in preventing damage to large, unprotected glass areas in heavy seas.

If exterior shutters add too much weight, or are too bulky to store in your available space, heavy, insulating cloths can be used instead. Waterproof vinyl or canvas weather cloths can be snapped onto the outside of the windows, or thick, quilted, fiberglass-filled cloths can be attached to the inside of the windows. These insulating cloths should be held tightly in place with elastic, velcro or snaps, to prevent warm air from escaping around them.

Heavy, foam-backed drapes can be used over the insulating cloths to provide further resistance to heat loss.

Many kinds of insulation are available which are suitable for use inside your boat, depending on the contours of the area, whether the area is visible, and whether it will be subjected to water incursion. Insulation with the highest possible R-value should be used under decks and on cabin overheads. Readily available acoustical ceiling panels can be used in many cases. Some of these panels are made of insulating fiberglass, can be installed with an air space above for further insulating value, and have a washable exterior surface. When installing insulation of this kind, be sure it is removable in areas that might require future access, such as below screws that hold handrails or tracks in place.

Interior surfaces of cabin sides can be lined with carpeting, cork or cork tiles. Latex or asphalt undercoating can be troweled onto the surface about 1/8-inch thick, or fiberglass batts and other more conventional-type insulating materials can be used in areas that are out of sight. In engine compartments, flame retardant fiberglass insulation with a reflective backing should be used.

Lockers and cabinets provide an air space between the hull and exposed cabin sides. These cabinets also should be lined with carpeting or other insulating materials. Everything stored in those lockers will provide either further insulation or thermal mass. Deep-pile bathroom carpeting is a good choice for lining lockers. It is made to withstand repeated washing and the deep pile provides many air spaces, therefore, much resistance to heat loss. It can be cut with ordinary household shears and can be stuck in place with double-faced tape in each corner for easy removal.

Fiberglass batts can be tacked underneath the cabin sole, and well-padded carpeting should be installed over the cabin sole. The carpeting and pad should be cut around access covers to the bilge and held in place with double-faced tape in each corner. Regular household carpeting in a synthetic fiber (nylon or polyester) is very satisfactory on a boat, dries quickly when wet, is unharmed by dampness, cleans easily, and has a long life.[1]

To prevent heat loss through closed hatches, they must be carefully weatherstripped and insulation should be added both inside and outside, if possible. The outside of hatch covers can be insulated by laminating them with wood strips.

This can be done very simply by cutting 1/4 inch X 2 inch or 3/8 inch X 2 inch teak strips into appropriate lengths. Arrange the strips on your hatch cover in a pleasing pattern and mark their location with a pencil. The strips can be numbered to facilitate getting them back in the same order. Then, glue them in place with epoxy glue and put weights on top to hold them in place until the glue dries thoroughly (about 24 hours). Bricks, concrete blocks, or stacks of canned goods make good weights. Remove the weights when the glue is dry.

If you want to use a black butyl rubber, or a caulking or bedding compound, between the slats, this is the time to do it. The sticking ability of the compound will be improved if you clean the area between the slats with acetone or denatured alcohol. Tape the slats carefully with masking tape on each side of the area to be caulked. This will be the most tedious part of the job. Apply the compound generously and press it down into the area between the slats with a small putty knife. Pull the tape before the compound dries, then step back and just admire your handy work for two or three days. It takes that long before the compound is dry enough for you to lightly sand and oil your newly laminated hatch cover.

Inside insulation for hatches will probably require the use of insulating cloths, again put in place with snaps or velcro.

Hatches leak enormous amounts of heat every time they are opened. The best way to eliminate or reduce this heat loss is to provide some kind of enclosure around the main hatch. A dodger helps a great deal, but a full cockpit enclosure is even better. You will want some kind of awning over the cockpit area, anyway, to keep off the rain and give you more usable space aboard your boat. If this awning is equipped with removable side curtains, it will form an efficient cockpit enclosure that will prevent cold winds from siphoning off

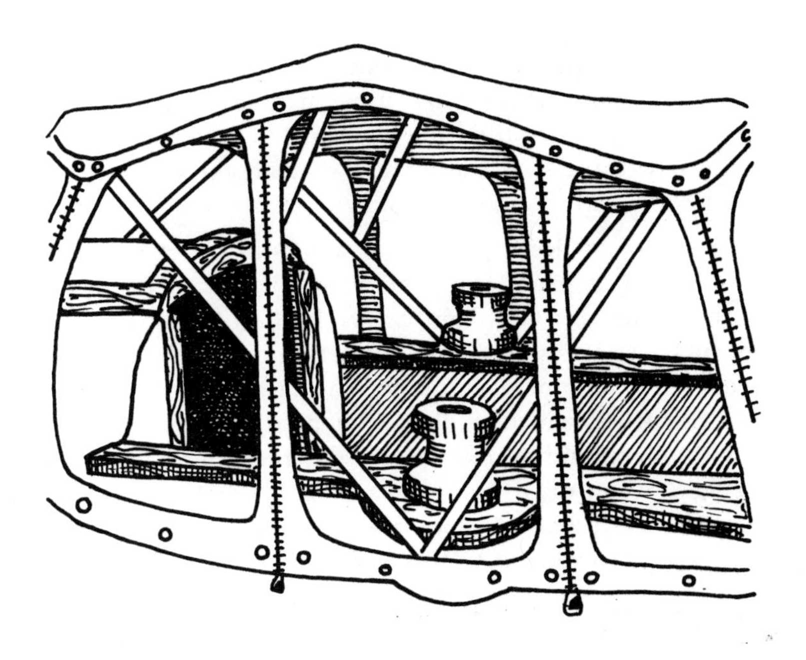

Canvas Bimini top with removeable plastic film side curtains

Aquatic Solar Greenhouse

your cabin heat. Also, if these side curtains are made of heavy plastic film, a greenhouse effect is set up and you will gain heat from the sun while protecting your entry from the wind.

While you have been creating a tight, dry boat, you also have turned your boat into an efficient *solar heat trap.* This means that it will make the best possible use of any heat or coolness it receives, letting that heat or coolness escape very slowly. The next step is to increase your boat's ability to collect solar energy and to store it.

To be a good *solar collector,* your boat must be able to let sunlight in when heat is needed and to keep it out when heat isn't desirable. It must also be able to let coolness in when needed.

The best way to use the sun's energy to heat your boat is to allow it to penetrate directly through the decks, hull, cabin top and windows. You can collect the maximum energy possible in cold weather by using dark colors, providing plenty of glass areas to receive the sun's rays, and by proper orientation to the sun. During hot weather, the use of shading and increased ventilation should moderate temperatures.

The color of decks, cabin tops and hulls strongly affects how much heat penetrates your boat, since dark colors absorb more sunlight than light colors do. Color is particularly important in areas where little insulation is used, but has less effect as the amount of insulation increases. This means that if your boat has a black hull and you plan to berth her in the tropics, you should do one of two things: either repaint her in a light color, or, if you have a special fondness for black hulls, heavily insulate the interior of the boat. By the same token, light-colored boats that make their permanent home in cooler climates either should be heavily insulated or repainted in a darker color.

Teak decks are marvelous heat collectors. This is of particular advantage in northern climates, but they needn't be ruled out entirely for use on southern boats. They can be heavily oiled in the winter, which will make them collect and hold the maximum possible amount of solar energy. Then, during hot weather, they can either be allowed to bleach out, or can be sheltered from the direct rays of the sun with a large harbor awning.

When it is possible, situate your boat ideally at dockside or at anchor. When this is not possible, you can compensate for a less than ideal location in relation to the sun by providing as many opening glass or clear plastic areas as possible.

The single most significant factor in capturing heat from the sun is the size and placement of glass areas. On a houseboat or a yacht used primarily in fair weather, or for short trips, large opening windows or sliding glass doors can make an important contribution toward heating and cooling, if the openings are well-sealed and exterior shutters are provided. Extremely large glass areas are not practical for a cruising boat, however, because of the danger of breakage in heavy winds and heavy seas. Therefore, the cruising boat should have many *small* opening windows or ports.

Besides reducing the amount of electricity used for lighting, glass areas capture heat through the "greenhouse effect" as explained earlier. Sunlight passes through the glass and is changed into heat as it hits an interior surface. Most of this heat is trapped in the glass and the interior spaces of the boat, and very little escapes into the atmosphere.

The amount of solar energy that passes through a south window on a *sunny winter day* is more than the amount that passes through that same window on a *sunny summer day,* for several reasons. The sun shines directly on a south window for *more hours* in winter than in summer. Also, the sun's rays strike windows at a more perpendicular angle in winter than summer because the sun is closer to the southern horizon in winter. Therefore, it is important that you have glass or transparent plastic areas on all four sides of your boat so that, regardless of its orientation to the sun, you will always have a south-facing transparent area.

Ordinarily, there will be a row of windows or ports on each side of the cabin. If any of these windows are fixed panes, they should be replaced with opening ports. The bow and the stern of most houseboats and motor yachts have large glass areas, but sailboats seldom do. This can be remedied very easily.

A flexible, transparent plastic cover can be made to fit the open forward hatch and snapped into place. This will funnel sunlight and heat into the forward part of the boat via the "greenhouse effect." A cover to fit the open hatch could also be cut from hard, clear acrylic plastic and pressed into place with a gasket of foam weatherstripping. Aluminum foil or any reflective film can be glued inside the hatch cover to gather and reflect additional sunlight into the boat. The plastic cover can be put in place for use throughout the winter months. It should fit so that the hatch cover can close over it when the sun goes down each day, then can be reopened at sunrise.

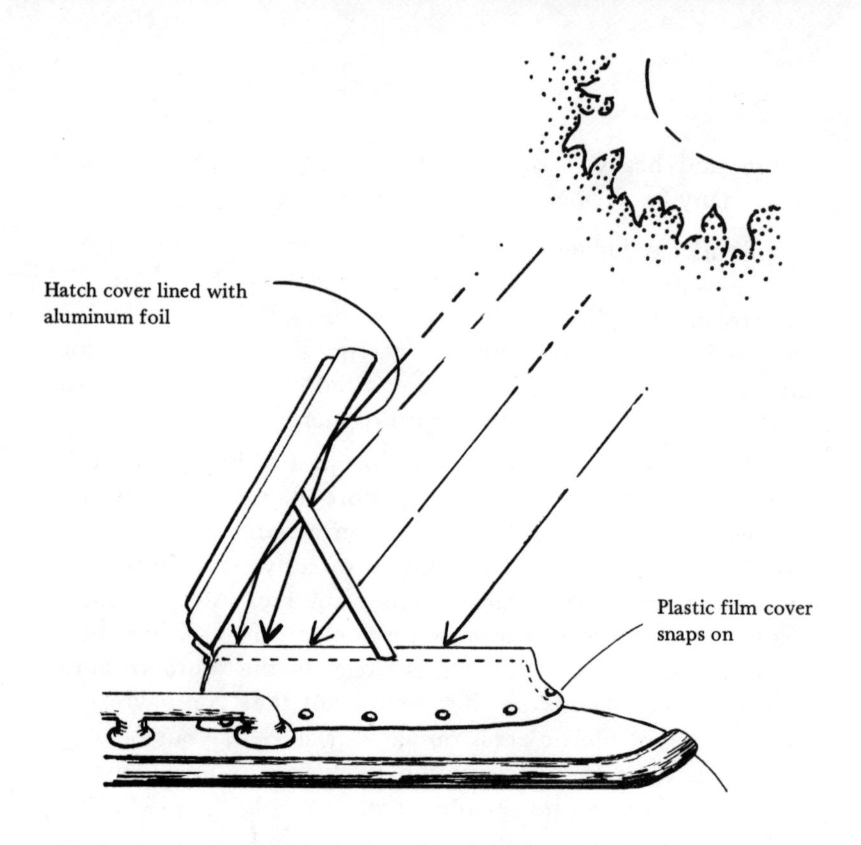

Solar Heater for Forward Hatch

Some kind of protective enclosure is an absolute necessity around the main hatch. Whether you use a dodger, a boom tent, awning or bimini, it should be equipped with removable side curtains and large areas of clear plastic should be sewed in. The clear areas will act like a greenhouse, again capturing heat, and will provide the final transparent area for the fourth side of your boat. All these transparent areas, when closed, will collect heat to warm your boat in the winter, but must either roll up or be removable to prevent the collection of heat when it is not desirable. Examples of this greenhouse-type enclosure are on pages 99 and 102-105.

Now that your boat has been turned into an efficient collector of heat, a few more modifications are required to make it an equally efficient collector of *cool.*

As previously mentioned, you should be able to open all windows and ports on the solar boat. To compensate for a less than perfect orientation to the sun, when cooling your boat it will be necessary to have openings on all four sides. You must also be able to shade these openings from the direct rays of the sun when heat gain is not desirable.

If you wish to use reflective film on unprotected glass areas, be sure to keep in mind that, while reflective glass will reduce heat gain to about 58% of the heat received in summer, it reduces heat gain in the winter, too, when heat gain is desirable, to only 19% of the available heat.

If the glass areas you wish to protect are actually acrylic plastic, as many yacht windows are, one alternate would be to apply the reflective film in the summer and peel it off in the winter. Although these films were not designed to be used in that manner, most of them will not adhere permanently to acrylic plastic. ("Scotchtint" brand sun Control Film A-20 is the one exception to that rule. It was designed specifically for use on acrylic glazing.[2]) Reflective film applied properly to glass areas become a permanent addition to your boat, so be certain that is what you want before applying film to glass.

The ideal answer to heat gain on unprotected glass areas is to install double panes of clear glass with an air space between. This will reduce heat gain in the summer to 83% with-

Boom Tent with Snap-On Side Curtains

out reducing winter heat gain any more than with a single pane of clear glass. This arrangement also reduces the escape of heat through closed windows in winter. A further refinement would be to line your insulating cloths with a reflective-type fabric. In this way, if the sun must shine on the window in summer, it can be closed and the insulating cloth can be put in place with the reflective side against the glass, further reducing heat gain. In winter, the reflective side can face the interior of the cabin after sundown, to reflect collected heat back in. Wind-n-Sun Shield, Inc., 1534 Avocado Ave., P.O. Box 1434, Melrose, Florida 32935, sells a fabric suitable for this purpose. It is a metalized polyester-backed white vinyl. You will also be able to find lightweight reflective-type overcoat lining materials at most large fabric outlets.

Your biggest weapons in combating unwanted heat gain in summer are shading devices. The awning or bimini you have installed over your cockpit area has been equipped with side curtains with transparent areas for winter heat collection. In summer, these side curtains can be replaced with

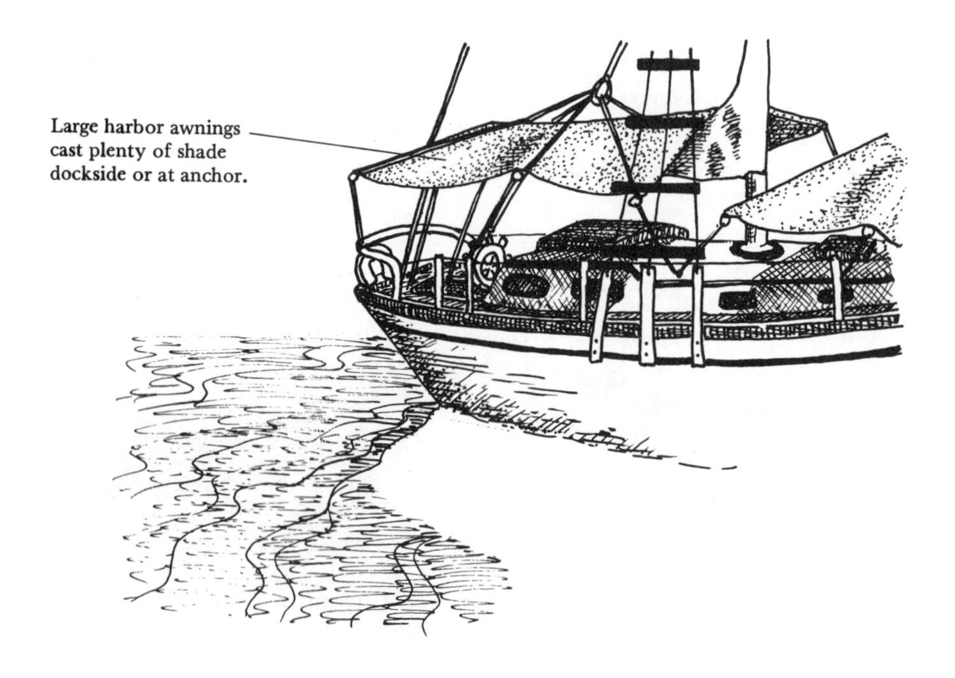

Boom gallows supports awning with snap-on side curtains

dark-colored, fine-meshed fiberglass screening. This screening will shade that area from all directions while allowing the passage of air, thus cooling the area. VIMCO Corp., P.O. Box 212, Laurel, Virginia 23060, sells "solar shields" which are aluminum-framed fiberglass screens in various ready-made sizes. These were designed for home use, but could be useful on certain boats.

Large harbor awnings that shade your entire boat are very successful in keeping your boat cool. Several smaller awnings can serve the same purpose. The important point is to shade as much of your boat as you possibly can. When making or purchasing an awning, select a fabric that is not transparent. Lightweight white canvas or dacron sailcloth is the poorest selection in fabrics because, although white does reflect sunlight, much sunlight (and heat) still passes through the cloth and causes a heat buildup on your boat. Vinyls or dark-colored acrylic canvases are excellent fabrics for awnings because they present a more opaque surface to the sun.

Use any means at your disposal to keep the sun at bay in the summertime. Be creative. Use roll-up shades, venetian blinds, shutters, eyebrows, aluminum foil shields, anything you can find that will prevent unwanted heat gain.

The next step will be to turn your boat into a solar storehouse.

Although the materials from which your boat is constructed are the greatest contributors to thermal mass, you can increase that thermal mass in certain ways. This will enable your boat to more easily moderate both daily and seasonal temperatures. Your boat will be able to hold heat during the winter and replace lost heat as the inside air cools. During the summer when cool night air or sea breezes are ventilated into the boat, the thermal mass of your boat will absorb any undesirable heat and prevent the inside air from becoming uncomfortably warm.

Hardwoods have more thermal mass than softwoods. So, when building lockers and cabinets, you can increase your boat's thermal mass best by using hard rather than soft woods. Asphalt tile also has great thermal mass and, although quite heavy, makes excellent flooring or even counter tops. Check the "Heat Storage Capacity Table" on page 7 for a guide to the best materials to add to your boat to increase its thermal mass.

You will note from the table that fresh water has a greater capacity to store heat than any other material shown, with salt water a close second. For this reason, the very water your boat floats in, as well as your own water storage tanks, can contribute much to your ability to store heat. Most primary water storage tanks are located in the bilge, which, since heat rises, will assist still more in heating your boat through natural convection. Additional bottles of water or canned goods can be stored under the floor boards to increase the capacity of your natural "heat storage bin."

The best way to evenly distribute the heat your boat collects is by natural thermosiphoning whenever possible. Air, when heated, expands, becomes lighter than the surrounding air and drifts upward. Then cooler air moves in to replace it.

This natural movement of the air can be assisted with dampers that you can either open or close securely.

In winter, the sun streaming through the windows on the sunny side of the boat will warm the air inside the boat. As this heated air warms, it will rise, displacing cooler air, which will flow down through the open floor level dampers. The cooler air will be warmed by the water in your natural heat storage bin and, once rewarmed, will drift upward again to mix with the air that has been warmed by the sun. Paint the top of water tanks with reflective metallic paint and install foil-backed insulation under floor boards to increase the effectiveness of this system. You should be able to establish a good thermosiphoning pattern if you have a large heat storage bin below the cabin sole, as shown in the diagram on the next page.

If you are unable to maintain good thermal circulation in certain areas – such as in lower level cabins – the warm air near the cabin top can be moved with the help of a simple, fan-assisted air circulator. This air circulator can be built into a hollow double bulkhead, or can be free-standing and portable. It consists of a 4-inch diameter PVC drain pipe long enough to reach from 4 inches below the cabin overhead to the top of a squirrel cage blower that is mounted in a box. The air circulator needn't be run continuously, but only long enough to set up a good thermal flow. When needed, it can be used to transfer heat from one cabin to another.

Natural thermal convection can work for you in the summertime, too, with the help of an open roof top ventilator or damper at a high point in the cabin, which will allow the escape of *unwanted* hot air.

To best use your passive solar heating system in *cold weather,* you should have all glass areas on the *shady side* of the boat shut tightly and covered with insulating cloths. Their drapes should be closed. On the *sunny side* of the boat, drapes should be opened and insulating cloths removed to allow full penetration of the sun's rays. Roof top ventilators should be closed. Ceiling and floor level dampers should be opened. In late afternoon – well before the sun goes down

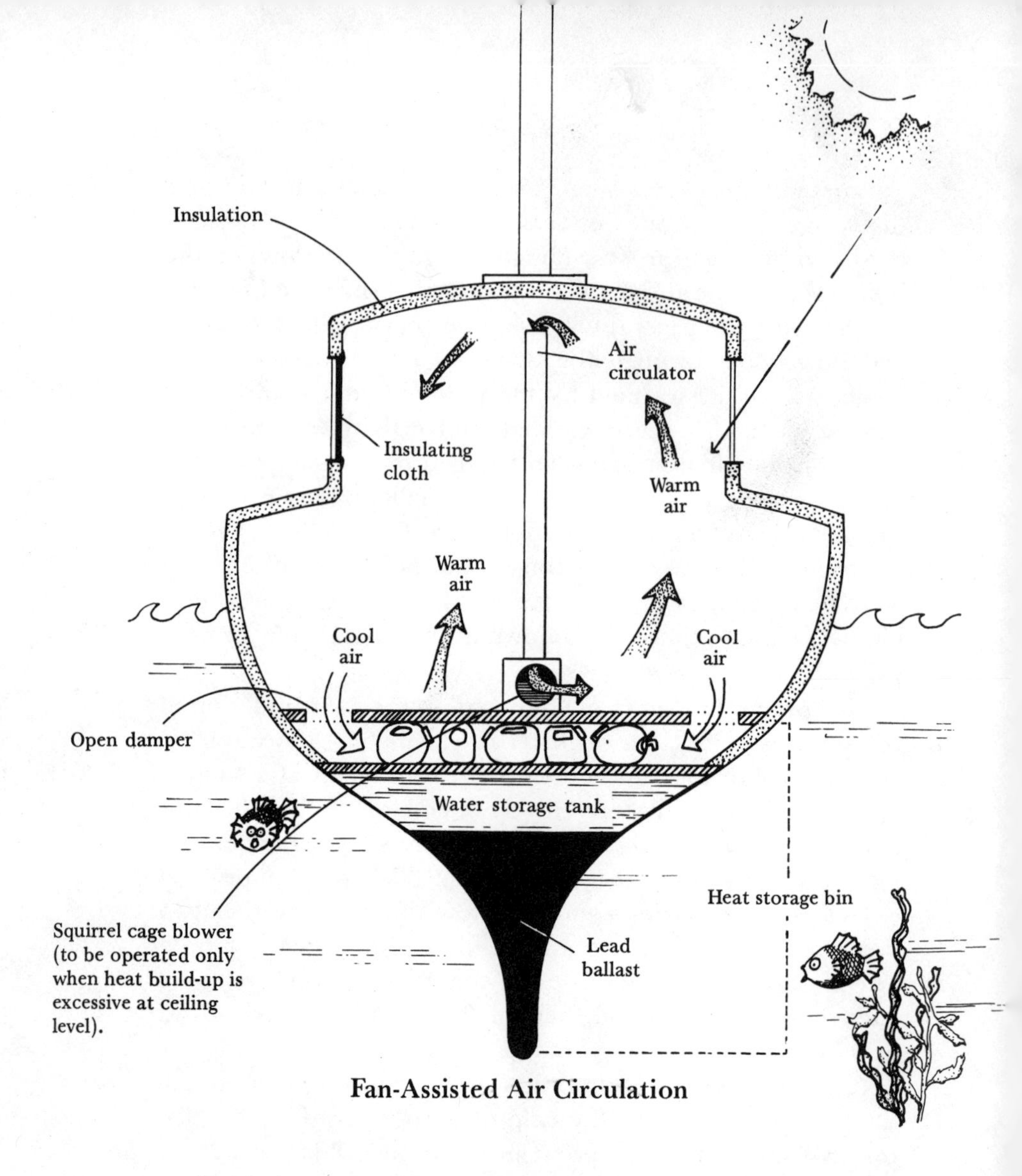

Fan-Assisted Air Circulation

– all windows should be covered to prevent the heat trapped inside from escaping.

In hot weather, reverse the procedure. First, awnings should be put in place, then all hatches and windows that are not exposed to the direct sun should be opened to allow the entry of every breath of a breeze. Roof top ventilators and all dampers should be opened. Any unprotected glass areas should be tightly shut, insulating cloths attached and drapes shut.

If you have followed the suggestions outlined in this chapter, you will find that the temperatures inside your boat remain much more constant and more nearly ideal than ever before, both seasonally and throughout the day and night. If, due to weight or space limitations, or for some other reason, you have not been able to make your boat as efficient a solar collector, storehouse and heat trap as it should be, and if your climate is very severe, some auxiliary heating or cooling devices may be necessary.

The most popular auxiliary solar heating device used in homes today is the window heating unit. This is a simple, easy-to-build, direct-heating unit that fits into the window of an existing structure. Several different designs are being used, and adapting one of these designs for use on your boat is a simple task.

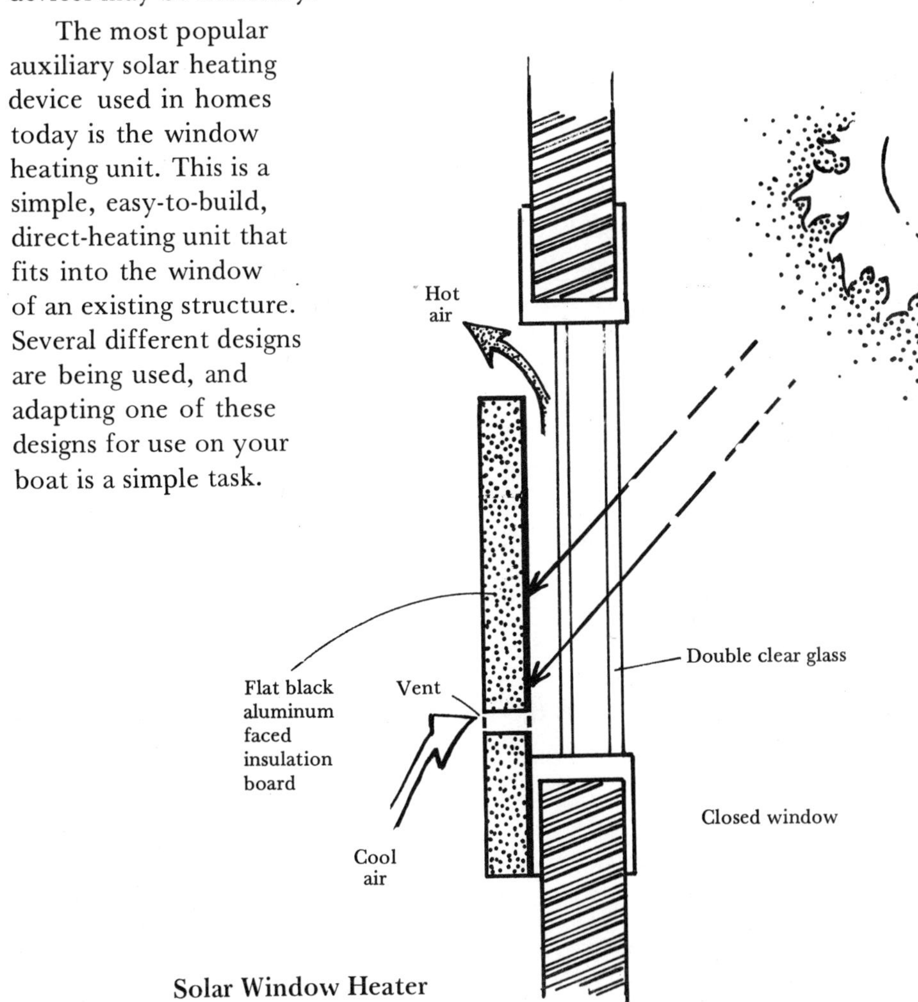

Solar Window Heater

Sunlight is absorbed by the glass-covered, blackened metal collector plate, heating the air inside the collector. The air is then circulated through the unit by natural convection and outlet temperatures on sunny days should be 100 degrees to 120 degrees F, regardless of the outside temperatures. This is a daytime-only heater, however, so the window must be tightly closed and insulated at night or on cloudy days to prevent the heater from working in reverse.[3] Plans for building a solar window heater for your boat are included in Chapter 11.

There is also a commercially built solar space heater for sale called the "Hot Shot" that is particularly well-suited for use aboard ship. It consists of a shielded tubular collector which mounts permanently on deck in a sunny spot and is ducted below by a small fan through a 4-inch pipe. A second 4-inch pipe takes out the cold air. Both pipes are flush to the bulkhead and have register coverings. The fan is offered as 110-volt AC or 12-volt DC with a solar cell option.[4] Although relatively expensive, this heater appears to be very efficient for daytime heating. A skilled home craftsman should be able to build a similar heater from salvaged materials at a great saving in price. If you want more information, write Solar Concepts, Box 263, Stafford, Texas 77477.

There is not enough heat storage space aboard most small boats to provide heat when the sun fails to shine for a long period of time. Therefore, to provide cloudy day insurance, consider installing a wood-burning cabin heater or fireplace as an auxiliary heater. There is nothing cozier than curling up in front of an open fire on a drab, snowy winter day, and you can feel secure in your knowledge that you are using a renewable fuel that is often free for the asking. Some sources for solid fuel stoves and fireplaces are:

Defender Industries, Inc.
255 Main St.
New Rochelle, NY 10801

Kristia Associates
343 Forest Ave.
Portland, ME
04204

or Kristia Associates
213 SW Ash St.
Suite 209
Portland, OR 97204

Jay Stuart Haft
8925 N. Tennyson Dr.
Milwaukee, WI 53217

Shipmate Stove Division
Richmond Ring Co.
Souderton, PA 18964

A wood-burning stove must be vented to the outside through a stove pipe with a Charley Noble above deck for heavy weather protection. By running this pipe through several cabins before venting it to the outside, you can heat a larger area with the same amount of fuel.

The proper installation and use of a wood-burning stove is an art in itself. If you are inexperienced in heating with wood, before purchasing, installing or using a wood stove, I suggest you read *Wood Stoves: How to Make and Use Them,* by Ole Wik. This is a thorough and comprehensive book and is available in paperback from Mother's Bookshelf, P.O. Box 70, Hendersonville, North Carolina 28739.

If you decide to install a wood stove for winter heating, paint the stove pipe above deck with flat black *stove* enamel. This will create a heat pipe and assist in *summer* cooling by siphoning the cabin's heat up through the stove pipe when the pipe damper is fully opened.

Instead of auxiliary heating, some auxiliary cooling may be required for boats in tropical waters. A boat at anchor will always be cooler than the boat at the dock, so head for the anchorage on uncomfortably warm days. If the boat remains uncomfortable, remember that a great deal of cool air can be funneled into your boat with some kind of wind chute. They are ready-made in several styles, some directional, some omnidirectional, but all work effectively if there is the slightest breeze.

You can easily make one of these wind chutes yourself from lightweight dacron sailcloth or from ordinary dacron lining material, which is available everywhere. Make a wind chute for each of your opening hatches. The more cool air you can deliver on a hot day, the better.

Another way to assist in cooling your boat is with water. When water evaporates into the air, the air is cooled, and the evaporation of 1½ gallons of water is equal to 1 ton of air conditioning.[5] Therefore, if you could wet the window screens through which air passes into your boat and keep them wet, the air would be cooled. The only problem with this system is that air passing through a wet screen on a hot day dries the

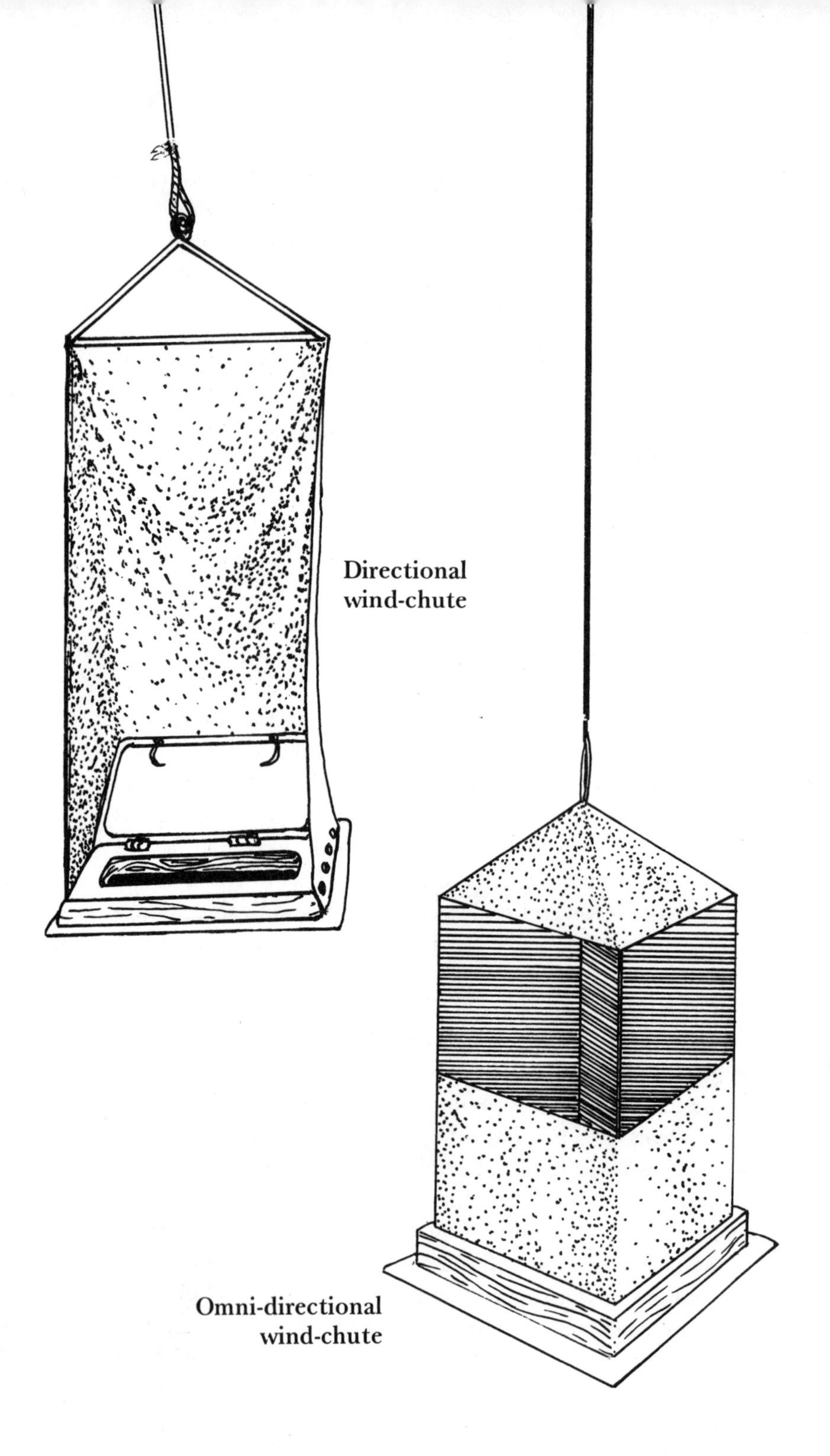
Directional
wind-chute
Omni-directional
wind-chute

screen out rapidly. Therefore, a more practical way of cooling by evaporation is to run a perforated garden hose down the center line of the boat over the top of the cabin and allow the water to trickle continuously over the cabin roof, down the sides of the cabin and off the deck.

In areas of high humidity, this system does not work. The air is already heavy with moisture and that moisture is already heavy with heat, so a method that dries the air will provide the most cooling power in humid climates. Best results are obtained by using numerous containers of dehumidifying chemicals, and by using low-amperage small fans to keep the cabin air circulating. Small cannisters of dehumidifying chemicals are sold at all large supermarkets in hot, humid areas. They all work well.

Excessive heat will seldom be a problem when underway, but if stronger measures are required to cool your boat at dockside, consider the use of large 110-volt AC fans before resorting to air conditioning. Fans use less electricity and you will have the benefit of fresh air and large open areas, rather than being sealed into a sterile, frigid, unnatural environment. A large floor model fan can be placed in a hatch to either exhaust hot air or blow fresh air into the cabin.

A 12-inch or 14-inch fan moves a great deal of air. *If* the boat is well-shaded, the cooling power of such a hatch-mounted fan should be sufficient for the hottest of days, especially if water is trickled over the cabin top. Air conditioning is seldom necessary.

A few solar air conditioners are on the market today, but they all require a much larger high-temperature solar collector than a small boat can accommodate. If you would like more information about solar air conditioning equipment, write to ARKLA Industries, Inc., P.O. Box 534, Evansville, Indiana 47704.

Now that you are comfortably afloat in your naturally-heated and cooled boat with all the comforts of home, including hot and cold running water, let's take a look at what the sun can do for you in the galley.

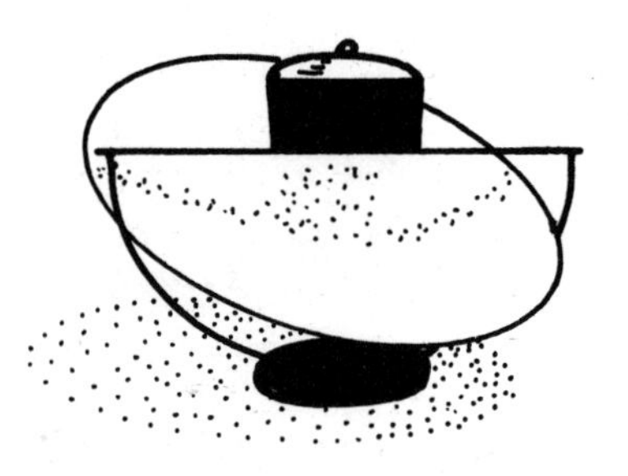

Indian Solar Cooker

9

Solar Food and Drink From Cookers to the Solar Still

A hot oven without a fire and a galley as cool as the outdoors are luxuries we all can have with the help of a solar cooker. Cooking food by boiling or baking with focused sunlight is one of the easiest ways to use solar energy. Much research has been done in the development of solar cookers, especially in low-income and rural areas, and in parts of the world where electricity and other fuels are in short supply or not available.

Although many clever designs have been introduced, there has been limited acceptance of the solar cooker among the very people who need it the most. The already-established cooking and eating habits of these people was the determining factor in deciding whether or not to use solar cookery. Most solar cooking cannot be done very early in the morning or in the evening, so it was acceptable only to those people who customarily ate their heavy meal at noon.

A small, inexpensive, lightweight, portable solar cooker is available from Davis Instruments, 857 Thornton St., San Leandro, California 94577. Food is skewered, wrapped in black-faced aluminum foil and placed at the focal point of the cooker (a parabolic collector). The small size of the collector limits the amount of heat it can produce, but this cooker works well when cooking hot dogs and certain quick-cooking kebobs, or when heating any food that is firm enough to be skewered.[1] I consider its light weight a disadvantage on a boat because, unless securely anchored, it takes very little wind to blow the cooker overboard.

Travis Brock of *Mother Earth News* put together a solar cooker that is almost identical to the one available through Davis Instruments, except for several important points. It is approximately twice the size of the Davis cooker, but still *small* enough to be portable. Built primarily of wood, it is heavy enough to stay put in the wind, but still *light* enough to be portable. Brock suggests the reflective surface be made from Flex-Sheet-Mirror, which provides a long-lasting surface that is easier to care for. Using all new materials, it costs only approximately $30.00 to build this cooker yourself.[2]

The VITA solar cooker is another simple, easy-to-build sun stove. It is rather large for use on most small boats, but it can be built from common materials – masonite, aluminum foil, wood, small strips of iron – for less than $10.00. The VITA Solar Cooker Construction Manual is available from Volunteers in Technical Assistance, 3706 Rhode Island Ave., Mt. Rainier, Maryland 20822, for a small fee.

In the *Scientific American* of June 5, *1878,* W. Adams of Bombay, India, described a solar cooker he built from wood and lined with mirrors. It was an eight-sided cone and was hinged onto a board so it could be moved to face the sun. The food to be cooked was placed in a blackened pail with a tight glass cover and hung at the center of the cone. This cooker was able to cook meat and vegetable rations for seven soldiers in less than two hours in January, the coldest month of the year in Bombay. The food could be stewed by adding water, or baked, if cooked without the addition of water.

Another sun-powered oven very similar to that 100-year-old design, called the "Solar Chef," is available commercially from the Solar Ray Co., Box 13573, Phoenix, Arizona 85002, for less than $200.00. The Solar Chef is portable and durable, and like most solar cookers, it costs nothing to operate. A typical test for this oven showed that it took only 20 minutes for the internal oven temperature to reach 300 degrees F on a sunny winter day when the temperature all around the cooker was 27 degrees F, and it cooked 4 lbs. of meat and vegetables in 70 minutes. The builders of the oven claim that it will cook up to 15 lbs. of food at temperatures up to 500 degrees F. Also, the actual baking area of the solar oven measures 36 inches wide X 18 inches deep (it is also an eight-sided cone shape), so it is large enough to bake bread or cookies.

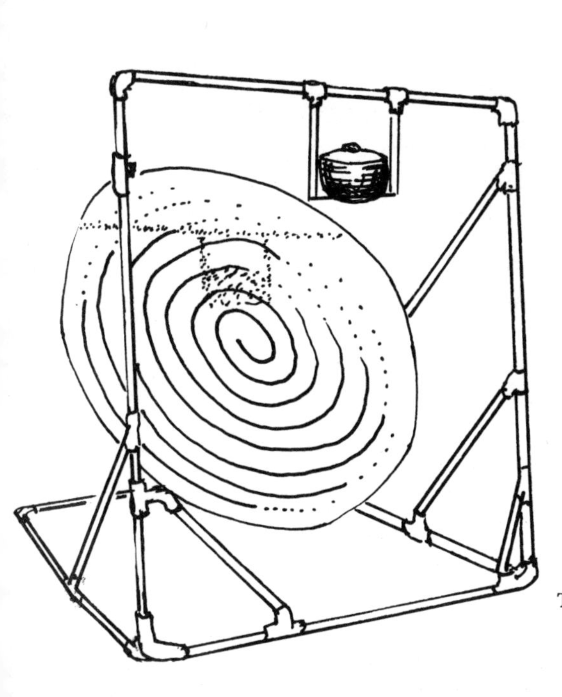

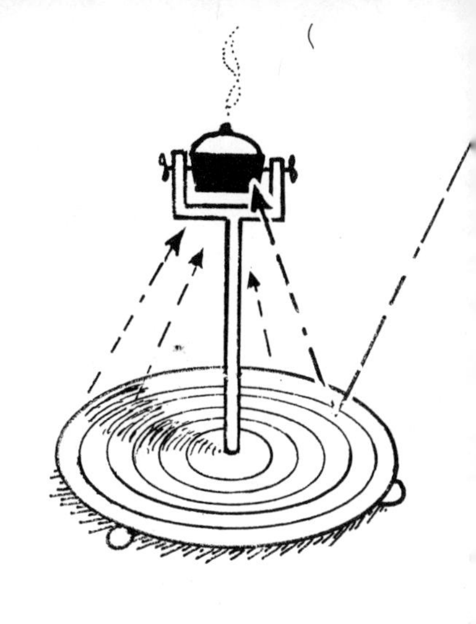

VITA
Solar Cookers

The sun reflects heat onto center of pot from all the rings at the same time.

This cooker is too large to be considered really stowable on a small boat when not in use, but it is so efficient that you might consider it worthy of a permanent location somewhere aboard your boat.[3]

Another type of solar cooker can be easily and inexpensively constructed using a Fresnel lens to focus the sun onto the food to be cooked. Inexpensive plastic Fresnel lenses are readily available and are powerful enough to boil water in less than three minutes. Construction details for a solar cooker of this type are shown in Chapter 11. This cooker is very convenient for making tea or coffee, and other foods can be prepared by steaming or boiling, or the cooker can be used like a double boiler. The only disadvantage to this cooker is the great care that must be exercised when focusing the lens. It is like a giant magnifying glass and is capable of producing temperatures up to 3,000 degrees F. Therefore, unless carefully handled, it could start a fire or give you a nasty burn. It is also imperative to wear dark glasses because the reflection and glare near the focal point could seriously damage your eyes.[4]

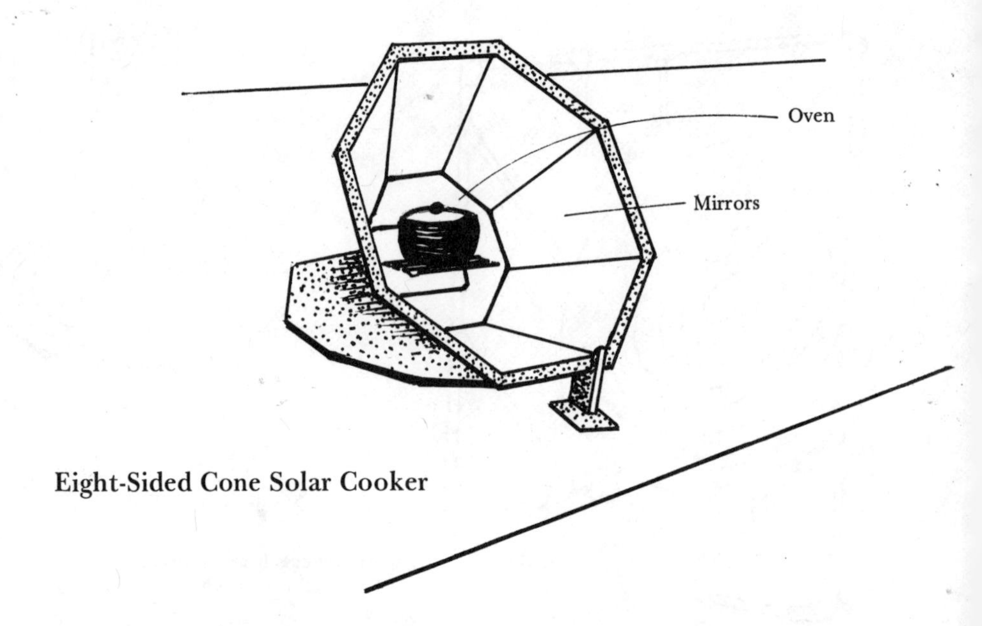

Eight-Sided Cone Solar Cooker

Several inexpensive solar ovens, kits and plans are available from Kerr Enterprises, Inc., P.O. Box 27417, Tempe, Arizona 85281, and from CLEVELAB, Inc., Box 2647, Littleton, Colorado 80161. Write for their catalogues.

All of the solar cookers mentioned in this chapter work quite well when the sun is shining. But what happens if the sun goes behind a cloud just as you are ready to prepare a meal – or worse, just *after* you have placed the food into the solar cooker? Unless the sun makes a hasty reappearance, you are back to more conventional cooking methods.

In order to overcome this obvious limitation, experiments have been underway for many years to develop a solar stove with a heat storage capacity that will allow for cooking at night and in cloudy weather.

Dr. M. Telkes has done a great deal of research in this area, using glauber salts as a heat storage medium, but the most promising development has been made by a University of Florida professor, Clayton A. Morrison, and his students. They are perfecting a solar stove that uses cottonseed oil in a closed system as a heat transfer medium. The oil is pumped from a high-temperature-concentrating-type collector into a

Fresnel Lens Solar Cooker

tubular surface unit at temperatures of 400 degrees F to 500 degrees F. They have added a tracking device to assure optimum focus on the collector at all times, and have provided for night cooking with a hot oil storage area. The tiny, low-drain electric pump is the only non-solar power used in the stove, and it could be operated by a solar cell array.[5]

The solar stove these Gainesville, Florida, students are working on operates very much like a stove used at the Mt. Wilson, California, Solar Observatory of the Smithsonian Institution, and described in a June *1923* issue of *Science and Invention.* The Smithsonian solar stove used engine oil as the heat transfer medium and cotton batting for insulation, and operated without an electric pump by placing both the cooking area and the oil storage tank above the collector. The hot oil circulated, therefore, by natural thermosiphoning.[6]

With the use of more modern materials, some of these older designs are very adaptable for use today.

Whichever type of solar cooker you choose, there are many ways you can enhance its cooking ability. Old fashioned black iron pots with pyrex lids are the best possible utensils to use in solar cookery. Although they do contribute a significant amount of weight as cargo, there are many advantages to using these pans. They absorb sunlight because they are black. They hold heat because they are iron, and a pyrex lid increases their cooking ability through the "greenhouse effect." They need a lot of oiling, but little washing, to keep them in good condition, so they help you save water.

Another way to increase the efficiency of your solar cooker is to anchor your boat fore and aft if possible when using your cooker away from the dock. All solar cookers require that you focus the sun on the food to be cooked, and a moving target will take longer to heat, therefore longer to cook. So it is important to keep the boat as nearly as possible in one position.

To reduce the cooking time of meat and potatoes, use a sizzle stick or heat pipe. When this heat pipe is inserted into food and then placed in a solar oven, it will transfer wasted heat to the inside of the food, thereby cooking it from the *inside* out as well as from the *outside* in. This reduces cooking time by 35% to 50%.[7]

Nylon roasting bags are excellent aids in solar cooking, too, because of the "greenhouse effect" created, as well as their ability to keep cooking juices contained – which will keep the reflective surfaces of your cooker clean and *bright.*

On very windy days, most solar cookers will be much slower to produce a hot meal. This can be prevented by using a sheet of clear acrylic as a wind shield. It is lightweight, inexpensive and easy to store, and you will find yourself using it again and again in solar experimentation.

Another way to reduce the time your solar cooker is in use, and to even cook at night, is with the assistance of an insulated box. To use the box, food is first brought to the boiling point or above (at least 212 degrees F) with your solar cooker. Then the food is quickly transferred to the insulated box, where it finishes cooking by retaining its own heat. Or the box can be used like a warming oven to keep fully cooked food warm and ready to eat at any time.

The only danger in this system is that bacteria can multiply or leave behind harmful toxins, unless the food is properly prepared. Therefore, use an accurate thermometer to check the temperature, and be sure to bring the food to the boiling point as quickly as possible.[8]

Construction details for an insulated box are shown in Chapter 11. However, if you want to test the theory before building such a box, you can easily improvise. Let's test a hot box with a batch of brown rice.

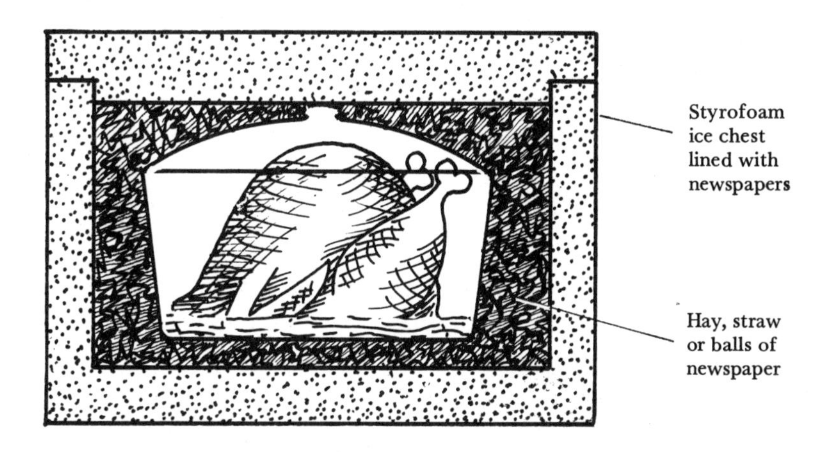

Insulated Box / Slow Cooker

Use an empty styrofoam ice chest, or any box with a tight-fitting lid that is large enough to accommodate your cooking pot. Line the chest with an old blanket or several layers of newspaper. Pack 2 or 3 inches of straw, hay or wadded-up balls of newspaper in the bottom of the box.

Put 2 cups brown rice in your cooking pot with 3 cups water. Bring to a boil with your Fresnel lens cooker. Boil for 5 minutes, remove from heat and place covered pot in center of the hot box, resting on bottom layer of packing.

Pack straw, hay or balls of newspaper tightly around the pot up to the brim. Put several layers of newspaper on top of the pot and place a pillow or a flotation cushion on top of the newspapers. Close the lid of the chest securely and take a swim, set out for the next port, or read a book. In 90 minutes, you may eat the most perfectly prepared brown rice you have ever tasted.

To prepare meat or vegetable stews, cut the food into tiny pieces and boil for at least 5 minutes using your Fresnel cooker, then transfer to the hot box for 3 to 5 hours to complete the cooking process.

Vegetable Solar Chowder

3 T. vegetable oil	3 T. flour
4 potatoes (with peeling)	4 C. water
4 carrots (with peeling)	1 sm. can whole kernel corn
2 stalks celery (with tops)	1 sm. can mushrooms (opt.)
2 onions	½ lb. grated cheddar cheese
1 green pepper	salt and pepper to taste

Cut scrubbed, fresh vegetables into 1½ inch chunks and brown in 3 T. hot oil in deep pot using Fresnel cooker. Stir in flour and cook 1 minute. Add water, then corn, mushrooms and seasoning. Stir well and bring to boiling point for 5 minutes. Transfer quickly to hot box. When ready to serve (3 to 5 hours), stir in ½ lb. grated cheddar cheese. Serves 4

You can put your dinner on to cook at noon, and serve it after the sun goes down. Breakfast cereals can be cooked the same way, later in the day, and placed in the hot box to be served the following morning.

Solar Oatmeal

½ C. soy flour	1½ C. chopped apple with skin (optional)
2¼ C. water	½ tsp. cinnamon
½ C. rolled oats	½ tsp. salt
½ C. raisins	

Add soy flour to water and bring to a boil with Fresnel cooker, stirring constantly. Slowly stir in rolled oats, then remaining ingredients. Return to boiling point for 5 minutes, then quickly transfer to hot box. Serves 4

DON'T FORGET TO WEAR SUNGLASSES AND GLOVES WHEN USING YOUR FRESNEL LENS COOKER!

Solar Drinks

Prepare a cup or glass of solar tea to go with your solar meals. This won't even require the use of your solar cooker. Just put 3 teabags into a clear glass jar with 4 cups of water. Place the jar in the bright sun and wait 30 minutes. Remove the teabags and pour. To cool the tea, tie a stout piece of line around the neck of the jar and hang it overboard (after the teabags have been removed) – or add ice if you must.

Ginger sun tea can be made the same way. In a glass jar containing 4 cups of water, add the following:

- 2 cinnamon sticks
- 10 whole cloves
- 10 black peppercorns
- 10 cardamom pods
- 4 slices ginger root (or 1 tsp. ginger powder)

Allow to steep in full sun for 1 hour. Strain out spices, add 4 T. honey, and cool overboard. Serves 4

Use your solar cooker every chance you get. Be adventurous; try new recipes and old favorites.

Solar cookery may not be an alternate to more conventional methods *every* day of the year, but it will open new horizons and is available to you every time the sun shines.

You've put the sun to work generating electricity, heating water, heating and cooling your boat, and preparing your meals. Yet, the sun has enough energy left for still more useful work. Sun drying foods is an excellent way of putting some more of that solar energy to good use.

People have been sun drying food to preserve it for thousands of years. Dehydration remains one of the most widely used methods of food preservation today for many reasons.

Drying preserves the vitamin, mineral and fiber content of foods better than preserving methods that require exposure to drastic temperature changes.

Dried foods can be stored in a much smaller area than frozen, canned or fresh foods – a definite advantage on a boat. Twenty pounds of tomatoes, when canned, fill 11 1-quart jars. The same tomatoes, when dried, fill about one 1-quart container and weigh just over 1 pound.

It costs nothing to dry foods; freezing and canning both require a healthy initial investment in equipment.

Dried foods, if kept dried, keep many months without refrigeration – another distinct advantage aboard ship.

You can stretch your food dollar by buying food in bulk when it is in season, then drying and storing it for later use.

Drying also answers the dilemma of what to do with the rest of that 60-lb. grouper after you've stuffed yourself with fresh fish and given away all you can.

Drying is definitely the answer when provisioning for a long cruise, and can also contribute to many healthful, enjoyable, delicious meals ashore.

Foods can be dried in a dehydrator, in your oven or in the sun. Instructions for building a small, exceptionally good food dryer that is small enough to use aboard a small boat, are included in Stella Andrassy's book, *The Solar Food Dryer Book,* published by Earth Books, 145 Palisade St., Dobbs Ferry, New York 10522.

If you feel a portable dehydrator would take up too much room aboard your boat, you can dry foods with virtually no equipment at all. All you need is a bright, sunny day – or several bright, sunny days in a row, depending on what food you are drying. And your galley stove can lend assistance if the sun should fail to shine.

The food to be dried is spread out on cookie sheets, aluminum foil, or butcher paper, or threaded on a stout piece of line, then placed in the sun or hung in the rigging. At sundown, the drying food must be brought inside. It can be stacked with layers of paper toweling or brown paper between the layers, then returned to the sunshine the following day.

The food should be spaced to allow for good air circulation, should be turned frequently, and returned to the sunshine each day until drying is complete. If the sun should fail you during the drying cycle, place the food in your oven and set the temperature at 100 to 140 degrees F. Prop the oven door open to allow for good air circulation. If you have no oven, prop the food on a rack 6 to 8 inches above the top burners, and set them at the lowest possible temperature. Turn the food frequently.

Once the food is completely dry, it should be packed loosely into ziplock bags, cans or jars, and placed in as dry, cool and dark a place as possible for storage. Aboard ship, containers of dried food should be inspected frequently for mould. If any foods show signs of increased moisture content (i.e., swelling), they can be spread in the sun again and re-dried. Mouldy food should be thrown out.

Any food that is fresh and fully ripe can be dried. When selecting food to be dried, however, remember that if it isn't good enough to eat fresh, it isn't suitable for drying. Foods that respond best to drying are listed below. (Foods that are not listed can be safely dried, but some flavor or texture may be sacrificed in the drying process.)

artichokes *
asparagus *
beans, green *
beets *
broccoli *
brussel sprouts *
cabbage *
carrots *
cauliflower *
celery *
corn on the cob *
corn, cut *
eggplant

horseradish
mushrooms
okra
onions
parsley
peas *
peppers
pimientos
potatoes *
spinach and
other greens *
squash *
tomatoes

apples *
apricots *
bananas
figs
grapes
mangoes
nectarines *
papayas
pears *
peaches *
persimmons
pineapple
prunes

Lean meat and fish

* Some authorities believe that these vegetables must be blanched and these fruits should be sulphured to prevent loss of color and flavor during the drying process. Others prefer the slight loss of color or flavor in unblanched foods to the slight loss of vitamins and minerals in blanched foods, another compromise for you to choose.

Foods to be dried should be selected carefully and washed. They can be sliced or left whole; the smaller the slices the quicker they will dry. Spinach and all other greens, green beans, apple rings, and many other foods can be strung on a stout cord with a sail or upholstery needle. When stringing food for drying, leave a space between the pieces to allow for air circulation and hang the strings from the rigging.

Parsley, mint, sweet basil, and other herbs and spices should be placed loosely in small brown paper bags and those bags should be clipped to the rigging with clothespins. Other fruits and vegetables can be spread into a single layer on cookie sheets, sheets of aluminum foil, or butcher paper, allowing space between the pieces for air circulation. The food should be turned several times a day when being dried on a flat surface.

Drying time may be one to seven days depending on the temperature, humidity, the size of the pieces of food, and the natural water content of the food.

A slightly different technique is required to prepare meat and fish for drying.

Select *lean* meat, remove all fat, and cut it with the grain of the meat into strips ¼ to ½ inches wide. Strips should be 4 to 8 inches long. Lay the strips out in a single layer on a cookie sheet or piece of aluminum foil. Brush with a marinade of garlic powder and liquid smoke, soy sauce, worcestershire or other steak sauce. Sprinkle *liberally* with salt and pepper. Turn the strips and repeat the process on the other side.

Place the strips, layer upon layer, in a large bowl or crock and place a weighted plate on top. Let stand for 6 to 12 hours; remove meat strips and blot dry with paper toweling. The meat must then be dried for 4 to 6 days.

An open barbecue grill is a good place to dry the meat. The bottom of the barbecue should be lined with aluminum foil to reflect the sun, and the strips of meat should be placed on the grill and suspended over the reflective foil. Turn the strips frequently. If you do not have a barbecue grill, thread the strips of meat on fishing line or other strong line with a button between the strips to prevent their touching. The meat is properly dry when it is leathery and chewy, but not yet brittle. At that point, it should be stored in air-tight containers, in a cool, dry place.[9]

It is difficult to preserve fish by air drying alone, especially if the fat content of the fish is more than 5%. Therefore, it is recommended that a technique combining salting and drying be used to ensure the best possible results.

Bleed the fish by cutting the throat the moment it is caught and remove the gills. Salting should be started as soon as possible after the catch.

Clean and scale the fish, remove the head *leaving* the hard, boney collar plates intact, and split it, removing the backbone. Scrub the fish with a stiff brush, inside and out, until it is scrupulously clean, leaving no blood, membrane, or viscera behind. Drain to remove excess moisture. If the fish is over 2 pounds, make diagonal cuts 4 inches apart to, but not through, the skin every place the fish is more than 3/4 inches thick.

The fish must be stacked on a flat surface that is slanted to allow the brine that forms to run off and be drained. A plastic dishpan makes a good container for dry salting fish. Prop a board up on one end inside the dishpan and stack the fish on the board, as shown in the sketch.

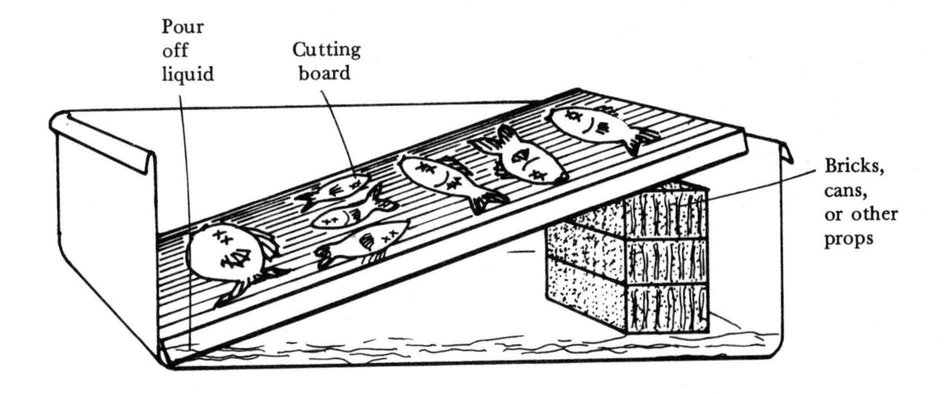

Drying Salted Fish in A Plastic Dishpan

Use pure, clean, fairly small-grain salt. Three-pound boxes of kosher or brining salt are available at most supermarkets. Do not use ice cream salt. Scatter a thin layer of salt on the board, first. Then dredge the fish in salt and arrange them on the board in layers, alternating heads and tails in each layer. Scatter a layer of salt between layers and stack the fish skin side *down* except for the last layer, which should be skin side *up*. Use about 1 pound of salt for each 4 pounds of fish.

Leave the fish in the salt for two or three days depending on the size of the fish. If the weather isn't suitable for sun drying, they can be left in the salt up to seven days while you wait for the sun to shine. Pour off the liquid that drains from the fish each day.

When you are ready to dry the fish, scrub it in clean salt water to remove all traces of salt or dirt, and drain for 20 minutes. Again, your barbecue grill makes a good drying rack, or if the boney collars of the fish are intact to prevent their falling apart, the fish may be threaded on stout cord or fishing line and hung in the rigging. However, to prevent oxidation (rusting), the drying fish should be shaded from the *direct* rays of the sun. If dried on a rack, the fish should be turned several times each day. When gathered up for storage each night, they should be stacked the same way they were stacked for salting – but without the salt – and a weight should be placed on top of the stack of fish. This helps press out additional moisture and shortens the drying time.

The fish should be dry in three to six days, depending on their size. Press the thick part of the flesh between thumb and forefinger. The fish is sufficiently dry if no impression is made. Wrap the cured fish in waxed paper and store in airtight containers in a cool, dry place. Inspect frequently and at the first sign of rust or mould, scrub fish in salt water and re-dry for a day or two.[10]

Now that you have a good supply of sun-dried fruits, vegetables, meat and fish, what do you do with them? Learning to use them well will be another new adventure.

Many dehydrated foods are even more flavorful and delicious than their fresh counterparts, and some don't even need to be reconstituted. Dried pineapples, mangoes and bananas, for instance, are sweet, chewy and fantastically good with no help at all. If you prefer, however, you can soften dried fruits by soaking in a small amount of cool water before eating.

Using dried vegetables is more of a challenge. Greens, cabbage and tomatoes need not be soaked. Simply add water to cover, and *simmer* until tender. Remember that high cooking temperatures will toughen dried foods. Root, stem and seed vegetables will require pre-soaking in cool water for 30 minutes to 2 hours. When soaking is complete and the vegetable appears to be reconstituted, simmer until tender. Be cautious when using dried onions, garlic or hot peppers. Remember that a *little* bit of dried onion is equal to an awful *lot* of fresh onion. I can't tell you exactly how much because it will depend on how small the pieces of onion were cut for drying and exactly how much moisture was removed in the drying process. Just heed the warning and proceed with caution.

Dried vegetables cannot be satisfactorily reconstituted for use in raw salads, and they are usually disappointing when served alone as side dishes, except for greens, cut corn and beans or peas. But dried tomatoes, onions, celery, peppers, mushrooms and many others are wonderful for concocting sauces, casseroles and soups, and provide much needed nutritious variety on long passages.

Dried beef jerky has become a popular snack. You may eat the dried meat as is, or use it in cooking. When cooking with dried meat, it should be wiped with a damp cloth before using. Soaking in cool water for 5 to 30 minutes will improve its texture when cooked. It can be creamed with a bit of dried onion and served over toast, noodles or rice. Dried meat also lends itself readily to curries or oriental stir-fry dishes. It can also be ground and mixed with various flavorings for a sandwich spread or party appetizer.

Dried, salted fish must be "freshened" before it is cooked, to remove excessive salt. Rinse and soak in cool water for 8 to 24 hours, changing the water several times. Then the fish may be broiled, fried, baked in milk, simmered, creamed or prepared in any of the ways previously mentioned for preparing dried meats.

If you would like more information on drying foods, one of the best books on the subject is *Stocking Up: How to Preserve the Foods You Grow Naturally,* by Organic Gardening and Farming, Rodale Press, 1973, and available at most first-class bookstores. Also, if you feel that you have enough space to accommodate a larger food dryer aboard your boat, plans for building such a dryer are available for a small fee from the Community Environmental Council, Solar Research Group, 109E De La Guerra, Santa Barbara, California 93101. Or request U.S. Department of Agriculture Home and Garden Bulletin No. 217, dated January 1977, *Drying Foods at Home.* This bulletin includes plans for both an electric food dryer and a natural draft vegetable dryer.

Drying foods in advance can enable you to dine in gourmet fashion as well as to provide yourself all the vitamins, minerals, fiber and protein necessary to maintain good health on the longest of cruises. And, in addition, in cases of emergency, the sun can help provide you with that other necessity – water.

Solar stills have been used for many years to convert sea or brackish water into usable fresh water. In 1872, in the desert of northern Chile, a solar still covering 51,000 square feet of land was built to provide up to 6,000 gallons of fresh water per day from salt water for use at a nitrate mine. This solar still operated effectively for 40 years until the mine was exhausted and the water no longer needed.[11]

Most stills are long, narrow, shallow black trays that hold the salt water. They are covered with an A-frame glass roof with sides sloping down to a trough on each side of the tray. The sun's rays pass through the glass roof and are absorbed by the 1-inch layer of salt water in the blackened tray. As the salt water heats, it vaporizes and is condensed on the underside of the glass roof, where it runs down into the troughs and flows into a reservoir. This is another example of the "greenhouse effect."

The amount of fresh water produced varies, but is ordinarily very small in relation to the space occupied by the commercial solar still. However, a small solar still aboard your life raft must be considered an indispensable survival tool. Small portable stills are available at army and navy surplus stores and through some marine supply outlets.

If you do not have a solar still aboard your boat, you may improvise by putting 1 inch of sea water into a deep pan. The pan should be a dark color or can be lined with a black plastic bag. Set a small container in the center of the salt water and cover the pan with a sheet of clear plastic. Put a fishing weight or a small stone in the center of the plastic so that it sags over the open container, as shown in the sketch. Water is more important than food in a survival situation. If you have no water, don't eat, or eat only foods with a high moisture content.

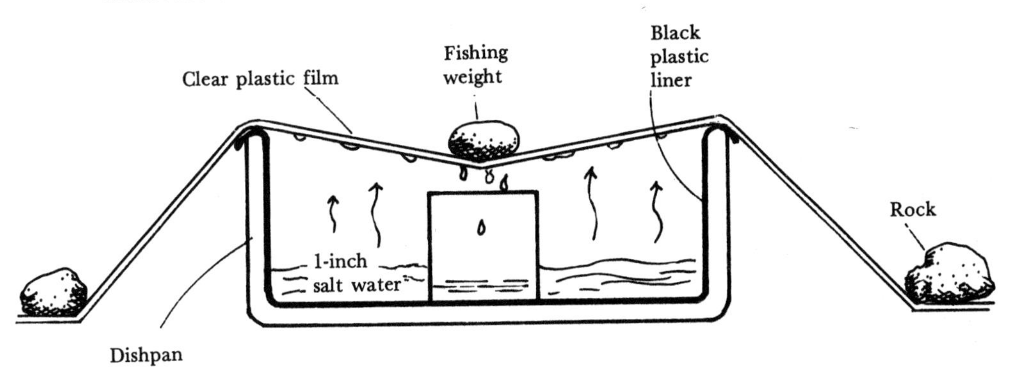

Improvised Solar Still

If you have been put ashore without water, 2 or 3 pints of drinking water a day can be extracted from soil or plant material with the help of a solar still. First, dig a hole about 40 inches wide with sloping sides. The depth in the center should be about 18 to 20 inches. A water container is placed in the bottom of the hole and cactus slices or other available plant material should be arranged on the sloping sides of the hole. Place a large sheet of plastic over the hole and weight

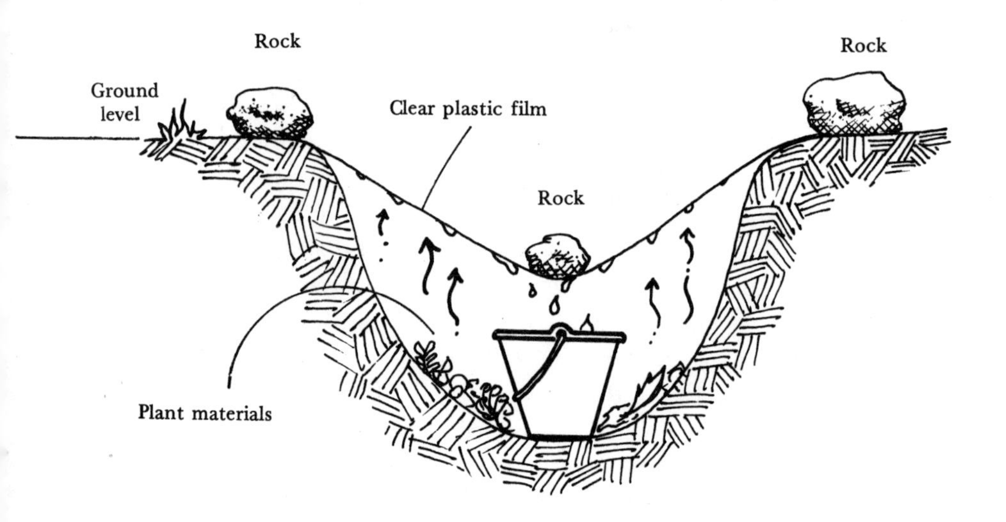

Survival Solar Still

the edges to hold it in place. Place a small stone or other weight in the center to form an inverted cone. Sunlight shining through the plastic will vaporize the moisture in the soil and plant material and after about one hour, droplets of fresh water will form and begin to run down into the container. [12]

10
Reducing Your Dockside Energy Demands

Being in tune with the *sun* is much like being in tune with the *sea.* You become aware of its tremendous power as well as its generous bounty. You learn quickly not to defy it, but rather to work in harmony with it. You would no more intentionally expose your body to the sun for an hour at high noon in July at the equator than you would intentionally sail into a turbulent, rock-strewn inlet on an ebb tide. And living in harmony with the sun is not just something to do when you're offshore. Solar energy can help you at the dock too, in many ways and for many reasons.

That fellow walking down the dock, who stops to ask you what that unusual device is on your deck, may be the next solar convert. He may be the person who will invent a new solar engine or he may be the one who will improve photovoltaic production techniques and reduce the cost of solar cells. Or you may be that person, yourself.

At one time, yachting was only for the privileged few, and yachts were kept at yacht clubs, while marinas were few and far between. Cruising boats, since they were equipped for offshore use with simple, limited 12-volt electrical systems, usually lay at a mooring. But the popularity of yachting grew, and with it the number of boats, then the number of marinas. Soon, the cruising boats joined the day sailors and runabouts at the dock, where they could plug into a dockside receptacle.

Unfortunately, most marinas were ill prepared for what happened next. Families began spending weekends at the dock. Some even moved aboard on a permanent basis. And, to avoid any comfort shock, along with these families came a steady stream of appliances: hot plates, toasters, heaters, blenders, irons, refrigerators, deep freezes, hair dryers, air conditioners, washers, dryers and the indispensable TV. As the number of appliances grew, the dockside voltage dropped, the boat-to-pier electrical cord got hotter and hotter and so did the wiring in the boat, creating a very real fire hazard. To further complicate the problem, the amount of electricity available varied from marina to marina, and often from receptacle to receptacle. Even today, there is no standardization of dockside electrical receptacles; therefore, it is necessary to carry an assortment of plugs, adapters and cords when cruising includes overnight marina stops.

The best way to avoid dockside electrical problems is to analyze exactly what your dockside power requirements are, then to be certain the wiring, both *in* and *to* your boat is adequate for these requirements. If it is not, then you must either rewire your boat, or reduce your requirements. The easiest way, of course, is to reduce your electrical demands and the sun is there, waiting to help you do just that.

Using the Electrical Consumption chart as a guide, list all the AC appliances you have aboard your boat. Complete the chart as you did for your DC appliances, following the instructions in Chapter 2. Instead of estimating the number of hours each appliance is used daily, however, enter the approximate time of day it will be in use in the "Hours Used Daily" column. Leave the "Daily Amp Hours Used" column blank. The purpose of your Electrical Consumption chart for AC appliances is to determine as closely as possible what your maximum ampere drain will be at any one time to prevent overloading the circuits.

A clamp-on type of ammeter can be used to determine the AC amperage of each of your appliances if that information is not either shown on the appliance itself or included in the instruction book that came with it. It is extremely helpful to mark the amperage of each appliance on the face of that appliance for future reference.

Once your Electrical Consumption chart is complete, add together the amperages of appliances you will probably be using simultaneously, such as hot plate, toaster and water heater. Or, what about air conditioner, hot plate and toaster?

If they add up to 15 amps or more, you may have a problem. If they add up to less than 15 amps, you are probably in good shape.

Most modern boats are equipped with 30-amp shore power wiring consisting of a 3-conductor, grounding-type shore power inlet receptacle with a water-tight cover, a main shore power disconnect, a 30-amp circuit breaker and standard 3-prong outlet receptacles mounted in the cabin, with suitably heavy wiring in between. The shore power cable that runs from the boat's inlet receptacle to the dockside receptacle for such a system should be at least 10-gauge or heavier to allow for voltage drop, and should be equipped with a locking, weather-tight, 3-prong, grounding-type 30-amp plug.[1]

The main problem, excluding the energy crunch, is that few marinas offer 30-amp, 3-conductor, grounded service, as you will discover when you try to plug into the dock receptacle. You must use an adapter that allows you to plug into the 15-amp, 3-conductor grounded dockside outlet that is commonly furnished, but you have now reduced your available amps from 30 to 15. So, keep that in mind and use your appliances discriminatingly. You may have to shut off the hot plate before using the toaster. Or, if necessary, you may bring in a second power cord from another dockside receptacle to operate an air conditioner or other heavy-demand appliance. If you have previously marked the amperage on the face of each appliance, appliance-juggling will be much easier.

A 115-voltmeter installed next to your master switch may bring you more shocking surprises about voltage variations in different marinas. The voltage may be well below normal in some of the older marinas. If it falls far below 115 volts, be extra cautious about using several appliances at once, as this low voltage could indicate inadequate wiring.

In some marinas, the electrical receptacles have no grounding conductor at all. They are simply regular household-type 2-prong outlets. In this case, to guard against both electrical shocks and galvanic corrosion, you must use a standard 3-hole ground adapter that fits into the 2-prong receptacle. A short grounding wire comes attached to the adapter; a

12-gauge wire, which is long enough to reach the nearest good ground you can find, must be attached to this short wire. Be sure *not to* use a fuel pipe, but look for another pipe or piece of metal that is in contact with the earth to which you may attach your grounding wire.[2]

In addition to the complications of having dockside electricity, some marinas charge extra for the use of this power. Some levy a large, flat fee; sometimes the electricity is metered. This is another persuasive reason to limit your use of dockside power.

If you have followed the suggestions in previous chapters, you have already lowered your power requirements. Solar cooking, solar water heating and solar space heating are even more successful at dockside than when underway because your boat is stationary and collectors can be set up in the optimum position. Instead of thinking of your solar power systems as alternates to electricity, it is important to think of solar as your primary system, with electricity, propane, oil or gas as alternates that you seldom use.

If a flat fee is added to your dockage bill for electricity, check out the economic feasibility of using only your own 12-volt power. Compare the cost of recharging your battery bank with the charges for dockside electricity. Your free fuel generators can be of some help. A hydro-generator will be useless to you *at the* dock unless you happen to be in an area, like some in the Bahamas, where a tidal current runs through the harbor at 6 knots or more. A wind generator will provide electricity at a dock, but wind speeds in protected harbor areas will be less than in most anchorages, which will mean an input of fewer amperes when at the pier. A solar generator will operate well wherever you are, as long as the sun shines, but you are not likely to have a large enough solar array to fill *all* your 12-volt requirements at *all* times. So, it may be necessary to run your engine at least for short periods several times a week, depending on what your demands for that period are. If the cost of fuel to run your engine (or your generator) for the required periods is more than the fee for dockside electricity, then *plug* in. That will be the time to re-examine your power requirements and explore new ways of reducing them. If you can afford to add solar cell arrays, by all means do so. Your initial investment will be high, but it can eventually pay for itself by eliminating electricity bills.

Individually metered dockside electricity is available at some marinas, which is especially good if there is no minimum charge – provided you have "solarized" your boat. Start a contest among owners of other boats of similar size to see who can pay the smallest electric bill without turning their lives into an endurance contest. It will be a perfect way for you to monitor yourself and your power demands, and you will discover that the "solar" boat uses very little electricity. And it isn't necessary to sacrifice comfort in order to conserve energy, either, if you remember that comfort is primarily a state of mind.

If, after careful study and reduction of your electrical needs, you still feel you need an electric heater, an air conditioner or refrigerator on board your boat, shop for the new appliance carefully. Study *its* power demands and weigh them against the availability of power and its ultimate contribution to your way of life. In the case of an air conditioner, for instance, is it worthwhile to you to carry around 70 lbs. or more of equipment 365 days a year in order to have the use of this appliance for a 4-to-6-week period in the summer? In the tropics the answer to that question may be a resounding "YES." But remember, the annual useful period of an air conditioner will depend more on exactly how well you have ventilated your boat and prepared it for hot weather, than on what your geographical location is. Weigh all the facts carefully before purchasing any new electrical equipment. It could be that another awning here, more insulation there, or an extra porthole or two might solve the problem better. If you do decide to purchase equipment of this sort, set up a maintenance schedule right away, following the manufacturer's recommendations. A properly maintained appliance will use less electricity than one that is neglected.

Small kitchen appliances, although convenient to have on land, occupy much needed space, add some undesirable weight and contribute little or nothing to galley cooking. A good quality Mouli grater, a wire whisk and a small chef's knife can do most anything those small appliances can do, while contributing no extra bulk or weight at all. Simplify in every way you can. It will free you to enjoy the pleasures of the sea.

11
Simple, Inexpensive Solar Projects

Solar Water Heaters

A 5-GALLON SOLAR WATER HEATER

In the interest of economy, the plans that follow use a 5-gallon metal jerry can as the combination collector-storage water heater tank. These cans can usually be found at stores selling genuine Army-Navy surplus for under $15.00. Any heavy-duty, leakproof, 5-gallon galvanized can will do the job, but a flattened shape is best for solar applications. A 6-gallon gasoline "saddle" tank would be appropriate.

If you prefer, a 5-gallon water heater can be custom built in any size appropriate to your space requirements (12 × 48 × 2 inches, 16 × 36 × 2 inches or any combination that occupies approximately 1155 cubic inches of space) in aluminum, galvanized, stainless steel or Monel for approximately $45 to $160, depending on the material and manufacturer. Consult your Yellow Pages under "Tank Fabricators" and get estimates from several firms, giving them the exact dimensions of the tank you want.

Whether you use a metal can, a saddle tank or a custom built water tank, by supplying an air vent, this system will gravity feed. Eliminate the air vent and the same system can be incorporated into your existing pressure water system.

1. Begin by washing the inside of the can or tank thoroughly with detergent and hot water to remove all traces of oil or gas residue. Flush several times with fresh water before continuing.

2. Lightly sand the bottom of the can to remove all traces of paint and to brighten the metal for soldering.

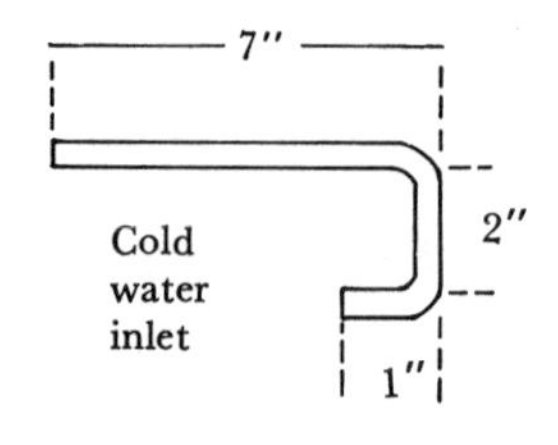

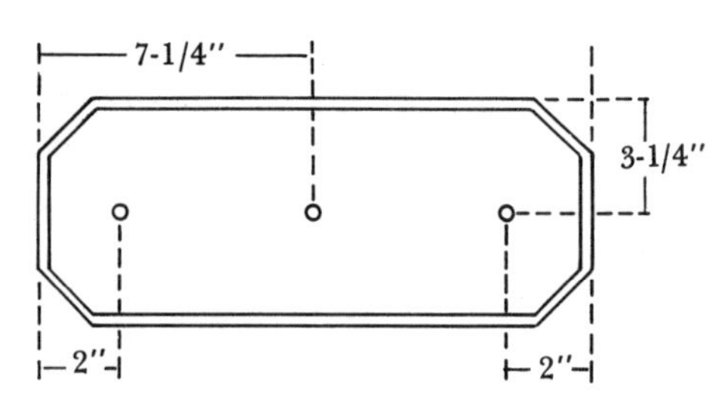

Bottom of can

Detail,
5-Gallon Can Solar Water Heater

3. Three lengths of 3/8-inch OD soft copper air conditioning tubing will be soldered to the bottom of the can: one 17 inches long; one 10 inches long; the third 5 inches long.

4. Bend the 10-inch length of tubing slowly and evenly, until it is in the shape shown in the sketch. Fill the tubing with sand when bending to prevent crimping.

5. Drill a ¼-inch hole each place a tube is to be installed. The correct location for these holes is shown in the sketch.

6. Drive a center punch into the three holes after they are drilled, gradually enlarging the holes until they will barely accept the tubing. The depression formed by flaring the holes inward will hold enough solder to make a strong bond.

7. Sand the short pieces of tubing in the area that will be soldered until they are bright, then work the pieces of tubing into the holes slowly and carefully, allowing 4 inches of tubing to protrude from the can.

8. Turn the can upside down so that the protruding tubes face upward and, using a torch, let the solder flow into the depression around the tubing. Use silver solder rather than lead solder to avoid contaminating the water in the tank and to achieve the strongest possible bond.

9. Cork the three open pieces of tubing; fill the can with water and check for leaks. If there are leaks, dry the can, re-solder and check again.

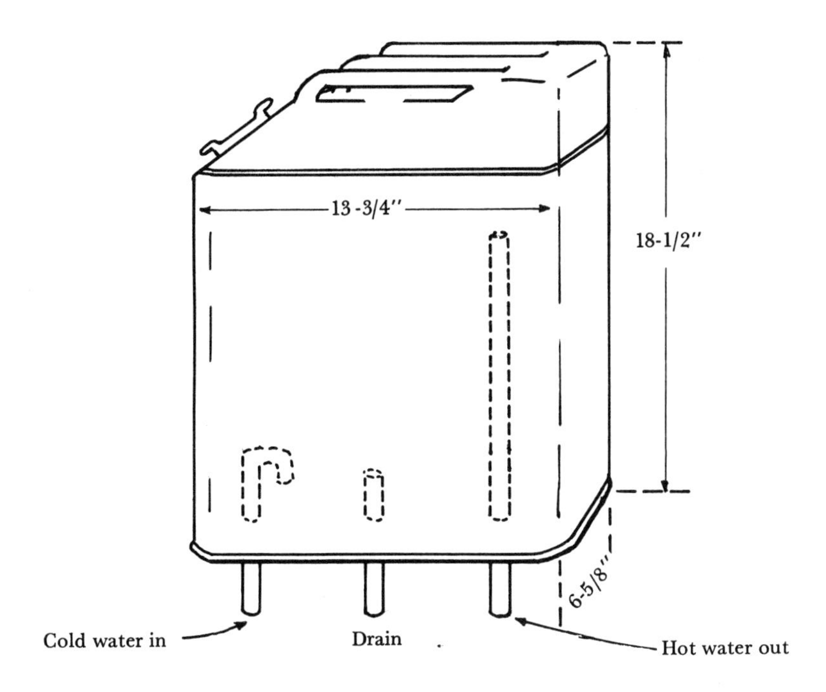

5-Gallon Can Solar Water Heater

A standard air vent may be installed in the top of the can at this time, or you may simply loosen or remove the cap for use in a gravity feed system. Do NOT install an air vent for use in a pressure system.

10. The can should now be lightly sanded, wiped first with acetone, then with full-strength household vinegar. When the can is thoroughly dry, give it at least two coats of flat black, *heat resistant* stove enamel.

The next step is to build an insulated box to hold your new hot water tank. The box should be built from ¼-inch exterior grade or marine plywood and 1-inch thick rigid insulating board: polyurethane, polystyrene or TF610 insulating board (which is aluminized on both sides). The dimensions shown in the patterns are for framing the metal jerry can shown in the sketch. If the can or tank you are using has different measurements, adjust the dimensions for the plywood and insulating board accordingly.

11. Cut all the pieces of plywood and insulation shown in the list of materials on page 145.

12. Assemble the back, bottom and two sides of the plywood box, using waterproof glue and brads.

13. Caulk the seams with waterproof caulking.

14. Using white glue, glue the previously cut insulation, first to the back, then the bottom, then the sides of the plywood box.

15. Drill three 3/8-inch holes through the bottom of the box and its insulated lining to accept the 4-inch pieces of tubing that protrude from the bottom of the metal can. Put the can in place, pushing the tubing through the holes.

16. Using galvanized strapping and bolts, secure the can to the box so that it cannot move in any direction.

17. Leaving 1¼-inch clearance on all four sides, glue the top piece of insulation to the top piece of plywood. Then slip the top piece into place after applying waterproof glue to the edges of the plywood. A brad near each corner of the plywood should hold the top in place.

18. Apply a heavy bead of waterproof caulking around the front of the box and carefully press the glass into place. When the caulking has completely set up, a strip of duct tape around the edges will seal and strengthen the joint.

19. Glue the front insulation to the front plywood leaving ¼-inch clearance on each side. The insulation should reach all the way from the top to the bottom of the plywood. Then, glue a piece of aluminum foil to the insulation using white glue. Slip the front into place and attach hinges with removable pins to the bottom of the front cover, and a hasp to the top of the front cover. This will enable you to either remove the cover entirely for collecting heat, or to leave the cover opened at an angle to bounce the rays of the sun onto the blackened water heater.

The box can be strengthened by applying galvanized flashing or strips of L-shaped aluminum to the exterior joints with epoxy glue and brads.

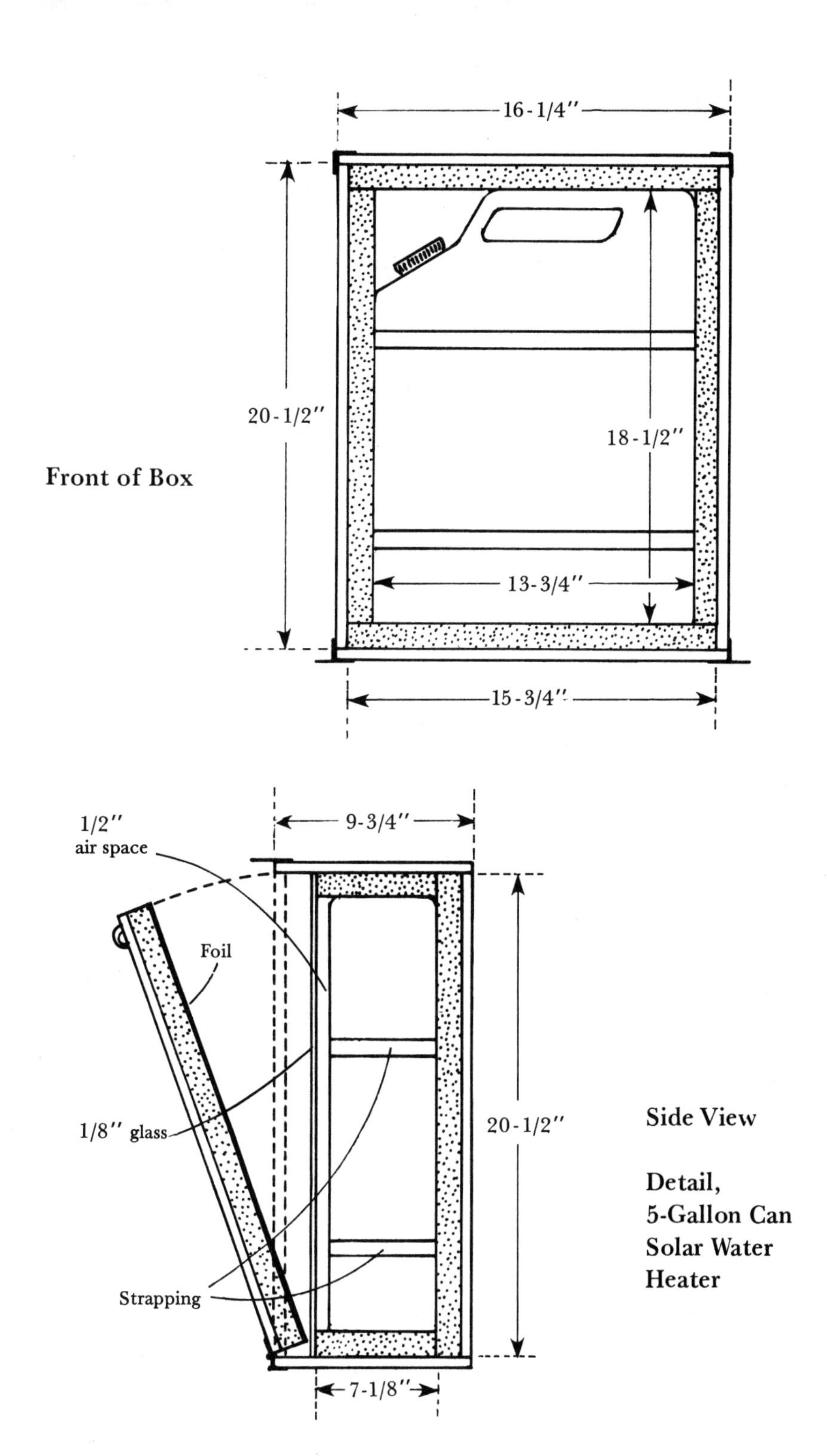

Side View

Detail,
5-Gallon Can
Solar Water
Heater

20. The box and cover should be sanded lightly and given at least two coats of exterior or marine enamel. Pay particular attention to sealing the edges of the plywood, especially if you have used exterior grade rather than marine plywood.

When mounting your solar water heater, drill three holes in your deck or cabin top that correspond to the protruding copper tubing. Cut the tubing to the proper length for the thickness of the deck and, after caulking heavily, push the tubing through and apply pressure fittings under the deck. The center piece of tubing should be capped. This is a drain plug, to be used when draining the tank. You may couple the cold water pipe to plastic tubing, but the hot water pipe must be coupled either to plastic or rubber *pressure* tubing, or to another length of copper tubing. The hot water line should also be wrapped in a thick layer of insulation.

Exactly how you incorporate the water heater into your water system depends on the design of your existing water system. Just remember to provide an air vent to gravity feed, and have an air-tight seal for a pressure system. The sketch on page 76 in Chapter 6 shows one type of gravity feed system. It shows one way to incorporate your new solar water heater into an existing pressure water system.

The water heater box should be attached to the deck or cabin top with through-bolted L-braces in each corner of the box. If your boat spends a lot of time dockside, try to locate your water heater so that the glass faces more or less in a southerly direction, next to the mast, on the deck or cabin top, or next to a stanchion on deck. When underway, your water heater's orientation to the sun will change, and its efficiency may be somewhat reduced, but it will still heat water – FREE.

To put the solar water heater into operation, remove the cover or open it to the best angle for receiving the rays of the sun. About one hour before sunset, close the cover securely, and the water will stay hot overnight.

List of Materials for
A 5-Gallon Solar Water Heater

one 5-gallon metal can
32 inches 3/8-inch OD soft copper air conditioning tubing
flat black heat resistant stove enamel

silver solder	duct tape
air vent	1 set hinges
epoxy glue	1 hasp
white glue	6 ft. flashing
3/4-inch brads	4 L-braces and bolts
30-inch galvanized strapping	exterior enamel
4 nuts and bolts	aluminum foil
waterproof caulking	tempered glass

You will also need a 4-ft. X 4-ft. X 1/4-inch piece of exterior grade or marine plywood, cut as follows:

2 pieces 16-1/4 X 20-1/2 inches (front and back)
2 pieces 9-1/4 X 20-1/2 inches (sides)
2 pieces 9-3/4 X 16-1/4 inches (top and bottom)

A 4-ft. X 4-ft. X 1-inch rigid insulating board should be cut as follows:

2 pieces 15-3/4 X 20-1/2 inches (front and back)
2 pieces 7-1/8 X 18-1/2 inches (sides)
2 pieces 7-1/8 X 15-3/4 inches (top and bottom)

The materials listed were selected to keep the size, weight and cost as low as possible, but substitutions may be made wherever desired. Use recycled materials every place you can. Acrylic glazing or two layers of plastic film can be substituted for the tempered glass. Make your own innovations, experiment a little, but get started today. A solar water heater will give you years of energy-free, trouble-free hot water. Let us know how *your* solar water heater works!

A THERMOSIPHONING SOLAR WATER HEATER

The following water heater incorporates a solar collector with a separate hot water storage tank. The storage tank is located above the collector so that thermosiphoning action will circulate the water naturally. Like the first water heater, the storage tank may be set up to gravity feed or be incorporated into an existing pressure water system.

Plans for this water heater also call for a 5-gallon metal jerry can, but this time we use it as a remote storage tank. The 18 X 48-inch collector should be located fairly close to the storage tank and at a lower level in order to ensure good thermosiphoning action.

We will begin by building the collector, which is a glass-faced insulated box, containing tubing fastened to a metal collecting plate.

1. Cut the plywood and rigid insulating board as shown in the list of materials. Assemble the plywood box using waterproof glue and brads. Glue the bottom and one set of

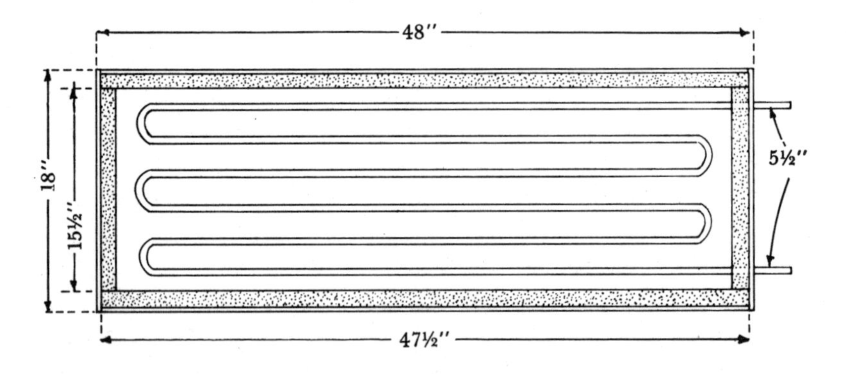

Top View, Detail of Thermosiphoning Solar Water Heater

side and end pieces of insulation into the box with white glue. Cover this area with aluminum foil and glue the foil into place shiny side out.

The next step is the assembly of the collector plate.

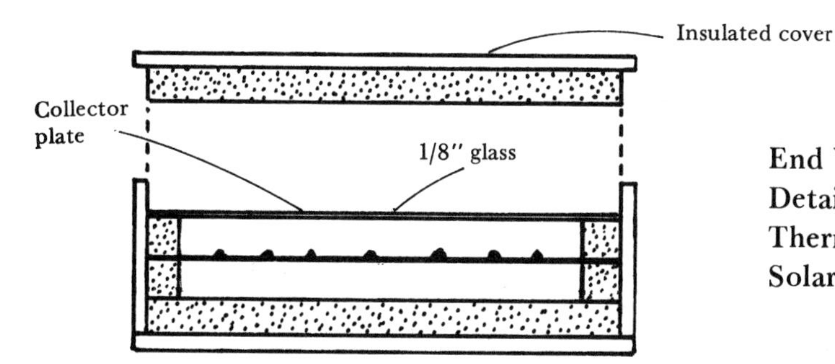

End View, Detail of Thermosiphoning Solar Water Heater

2. Using a marking pen, draw a diagram of where the copper tubing will lay (as shown in the sketch on page 146) directly on the sheet copper.

3. Straighten and true the coil of soft copper tubing, then bend the tubing slowly and carefully into the shape you have marked on the sheet copper. Note that the tubing will protrude 5½ inches beyond the copper sheet at the beginning and end of the tubing. Work on a flat surface, and take your time! Be sure that the tubing doesn't crimp and that it lies absolutely flat. It must make good contact with the sheet copper for good heat transfer.

4. Clean both the copper sheet and the tubing with emery cloth to ensure a good bond.

5. Lay the tubing in position and weight it to keep it in position. Using a torch, solder the tubing to the sheet copper every 6 inches. Heat from the soldering process will cause the copper sheet to curl, but it will straighten out again as it cools.

6. Give the entire assembly two coats of flat black, heat resistant stove enamel.

7. Mark the end of the plywood box where the tubing will pass through it, and drill two 3/8-inch holes.

8. Slip the collector plate into place with the tubing through these holes, and drill a small hole in each corner of the copper sheet through the insulation and plywood back. Bolt the collector plate in place through these holes.

9. Glue aluminum foil to the remaining insulation side and end pieces; then glue these pieces in place inside the plywood box. On these pieces, in a heavy bead of waterproof caulking, place the sheet of 1/8-inch tempered glass cut to

size. After the caulking is completely set, run a piece of duct tape around the edges of the glass to complete the seal.

10. Allowing ¼-inch clearance on all four sides, glue the cover insulation to the cover plywood. Then glue aluminum foil cut to size to the outside of the insulation. When using the cover, simply press it into place.

The corners of the collector box can be reinforced with flashing or pieces of L-shaped aluminum held in place with epoxy glue and brads.

The entire wooden box and cover should be lightly sanded and given at least two coats of exterior enamel before attaching it to the deck or cabin top with four through-bolted L-braces.

The next step is to turn a 5-gallon metal jerry can into a hot water storage tank.

Proceed exactly as indicated in the plans for the 5-gallon water heater. The only differences will be soldering two addi-

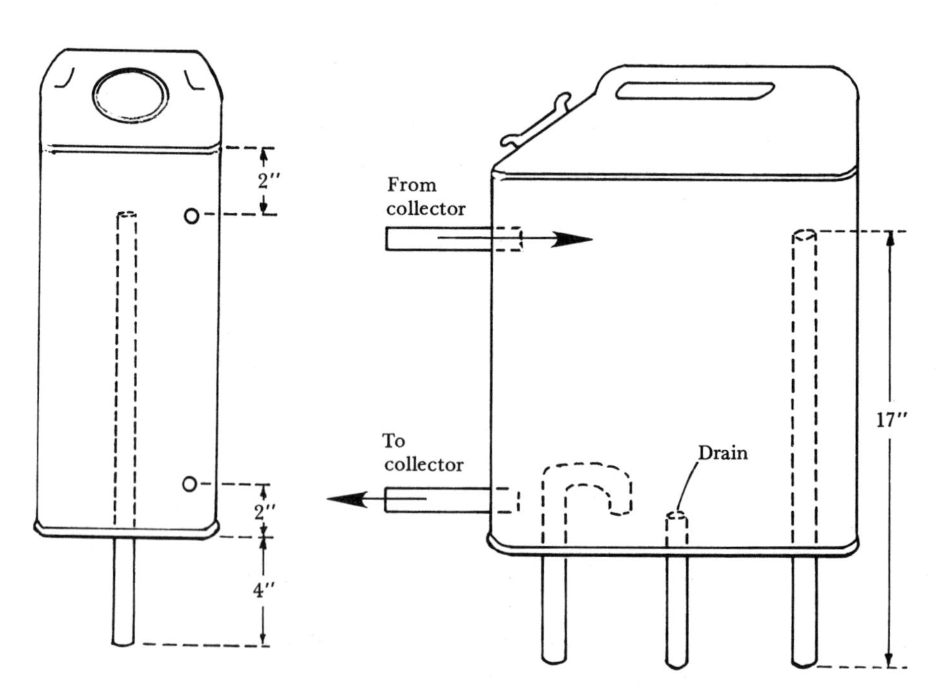

Hot Water Storage Tank

tional 5-inch pieces of tubing to the can opposite the side with the seam and drilling two 3/8-inch holes in the side of the insulated box to accept them. These two pieces of tubing allow the connection of pressure hose (or more copper tubing) between the collector and the storage tank.

The insulated box containing your storage tank will be attached to the deck or cabin top with L-braces, so that the bottom of the storage tank is level or higher with the top of the collector. Mounting details are shown in the schematic of the 5-gallon water heater.

After the storage tank and collector have been mounted, connect the two pieces of tubing that protrude from the collector to the two pieces of tubing that protrude from the side of the storage tank, using *pressure* hose or more copper tubing, wrapped heavily with insulation.

List of Materials for the Collector (for A Thermosiphoning Solar Water Heater)

sheet copper 17-1/2 X 47-1/2 inches
1/8-inch tempered glass 17-1/2 X 47-1/2 inches
3/4-inch brads
waterproof caulking
flat black stove enamel
33-ft. 3/8-inch OD copper tubing
flashing or L-shaped aluminum
aluminum foil
exterior enamel
duct tape
4 L-braces and bolts

4-ft. X 4-ft. X 1/4-inch exterior grade or marine plywood, cut as follows:
- 2 pieces 18 X 48 inches (cover and bottom)
- 2 pieces 4-3/16 X 47-1/2 inches (sides)
- 2 pieces 4-3/16 X 18 inches (ends)

4-ft. X 4-ft. X 1-inch rigid insulation board, cut as follows:
- 2 pieces 17-1/2 X 47-1/2 inches (cover and bottom)
- 4 pieces 1 X 47-1/2 inches (sides – 2 sets)
- 4 pieces 1 X 15-1/2 inches (end – 2 sets)

List of Materials for the Storage Tank
(for A Thermosiphoning Solar Water Heater)

1 5-gal. metal can	3/4-inch brads	6-ft. flashing
silver solder	4 L-braces and bolts	exterior enamel
epoxy glue	1 set hinges	aluminum foil
white glue	1 hasp	

42 inches of 3/8-inch OD soft copper air conditioning tubing

4-ft. X 4-ft. X 1/4-inch plywood, cut as follows:
- 2 pieces 16-1/4 X 20-1/2 inches (front and back)
- 2 pieces 8-5/8 X 20-1/2 inches (sides)
- 2 pieces 9-1/8 X 16-1/4 inches (top and bottom)

4-ft. X 4-ft. X 1-inch insulating board, cut as follows:
- 2 pieces 15-3/4 X 20-1/2 inches (front and back)
- 2 pieces 6-5/8 X 18-1/2 inches (sides)
- 2 pieces 6-5/8 X 15-3/4 inches (top and bottom)

Insulating Cloths

These insulating cloths are made from foam-backed drapery fabric, quilted cotton, and a space blanket. However, recycled materials can be used: any heavyweight, tightly woven drapery fabric, quilted cotton mattress pads or several layers of toweling, and reflective overcoat lining, for example. The cloths will be reversible, to be used with the shiny side *out* in hot weather and with the shiny side *in* in cold weather.

FOR PORTHOLES

1. Measure the height and width of the porthole and add 5 inches to each measurement. (If the porthole measures 6 X 13 inches, the fabric needed will measure 11 X 18 inches.)

2. Cut a rectangle that size from the drapery fabric and round the edges to match the porthole.

3. Cut a layer from the quilted cotton and a layer from the space blanket, using the oval piece of drapery fabric as a pattern.

4. Rack the three layers of fabric together every 3 inches, using dacron thread and an upholstery needle.

5. Stitch 1 inch from the edge, as shown in the sketch, and trim away the edges of the quilted cotton and the space blanket. Turn the drapery fabric in over the reflective layer. Turn under ¼ inches and stitch down, forming a casing for elastic, leaving a small space to pull the elastic through.

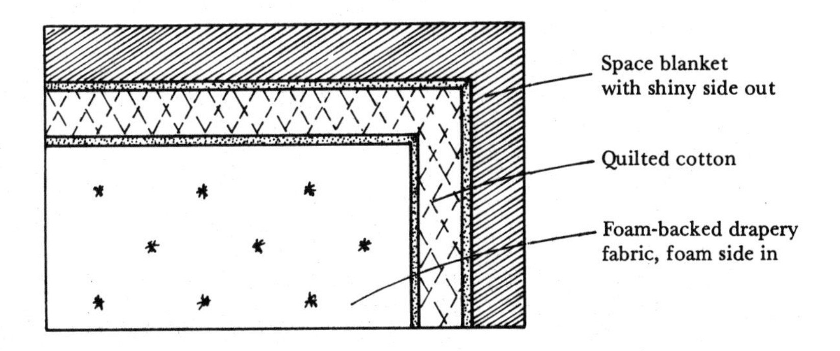

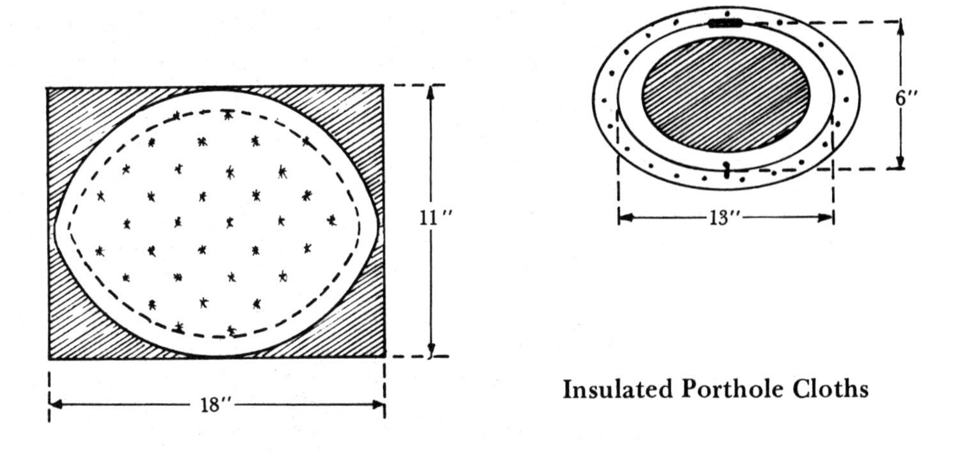

Insulated Porthole Cloths

6. Measure the perimeter of the porthole and cut a piece of ¼-inch elastic the same length.

7. Push the elastic through the casing and stitch the end together.

8. Slip the insulating cloth over the porthole to be sure it fits snugly around the porthole and tight against the cabin sides. If it does not, tighten the elastic and re-fit.

FOR WINDOWS AND HATCHES

1. Measure the length and width of windows and hatches, and cut a layer of quilted fabric and a layer of reflective fabric to size. The drapery fabric must measure 2 inches larger on each side, as shown.

2. Cut the drapery fabric, pin it in place and tack it to the other two pieces of fabric every 4 inches, using dacron thread and an upholstery needle.

3. Turn the edges of the drapery fabric under and stitch down at the edge of the reflective fabric, as shown.

4. Sew a strip of velcro around the edges of the insulating cloth on both sides in order to make it reversible.

5. Glue the other half of the velcro strip in place around the edge of your windows or hatches with contact cement. Velcro is expensive, but don't skimp. It is very important that the insulating cloths fit tightly around the window or hatch so that no stray air currents can escape.

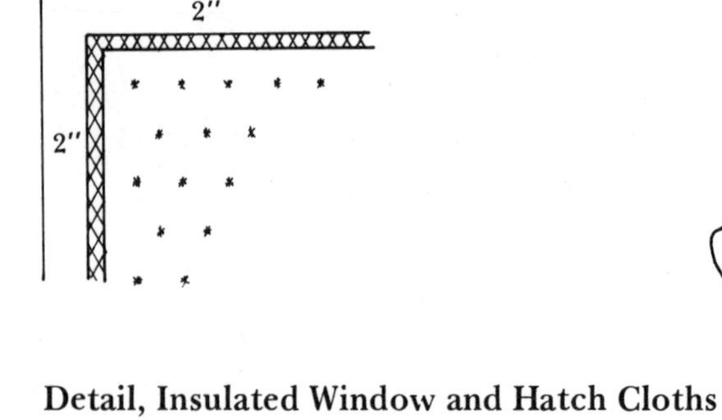

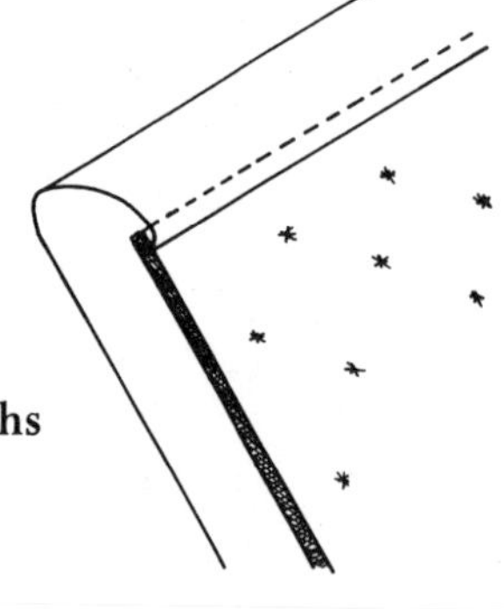

Detail, Insulated Window and Hatch Cloths

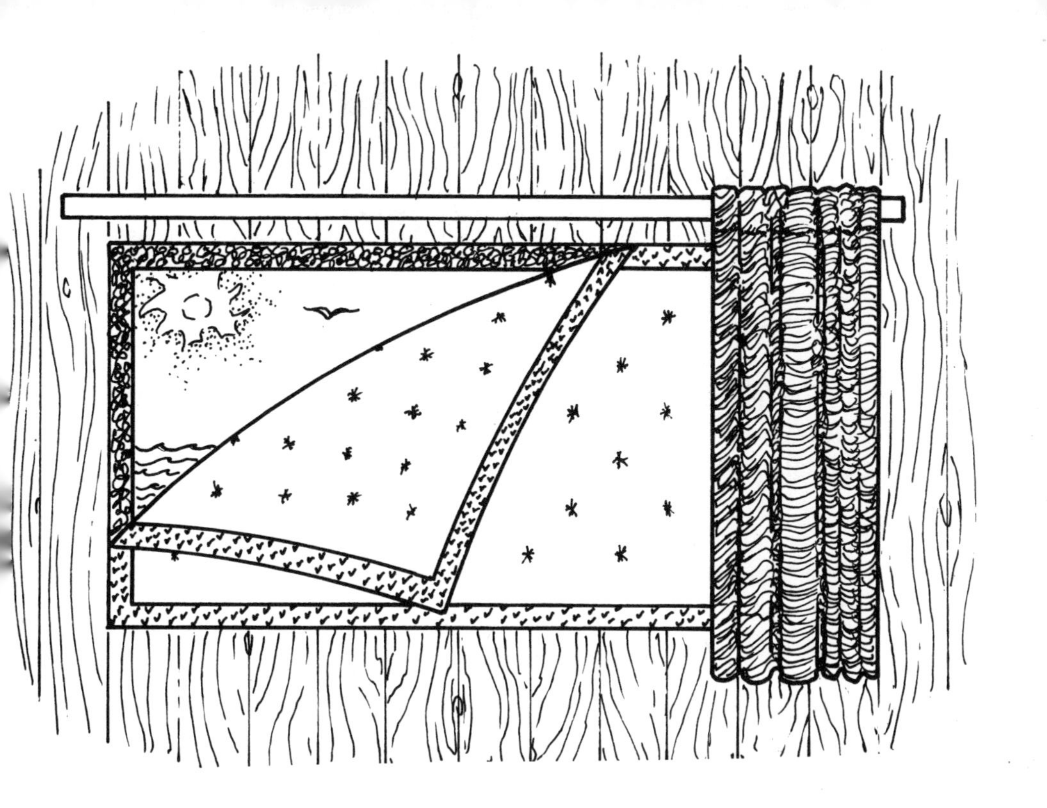

List of Materials for
Insulating Cloths

foam-backed drapery fabric*
quilted cotton*
upholstery or sail needle
reflective fabric*
dacron thread
Portholes: 1/4-inch elastic or shock cord**
Windows and hatches: 5/8-inch velcro** and contact cement

*There are 1296 square inches in one square yard of fabric. To determine the amount of fabric to buy, multiply the length plus 6 inches by the width plus 6 inches of each opening. Add the resulting figures of all the openings and divide the total by 1296. The answer will be the number of square yards of fabric you will need.

**Measure the perimeter of portholes, windows and hatches. For portholes, buy 1/4-inch wide elastic that same length. For windows and hatches, buy 5/8-inch wide velcro twice that length.

Space Heaters

FOR USE IN OPEN FORE HATCH

A great deal of heat can be captured below deck via the "greenhouse effect" by merely covering the hatch opening with glass or clear plastic. An open fore hatch makes an excellent solar heater.

1. Measure the hatch opening – length, width and depth – and cut a piece of heavy plastic film to size as shown in the diagram. The slit on one side is to accommodate the hatch adjuster; the slits at the top will accommodate the hinges.

2. Sew the corners together, using dacron thread and a sail or upholstery needle. The edges are not hemmed, but a small square of plastic film is moistened with water and stuck in position each place a snap is to be applied. When it dries, the square will stay in place.

3. Apply snap-type canvas fasteners where shown in the sketch.

4. Slip the plastic cover over the open hatch and mark each place a snap touches the cabin top, with a soft lead pencil. Studs will be screwed into the cabin top at these points. (These marine canvas fasteners are available from most marine hardware outlets. They are nickel-plated brass, and the studs are self-tapping. A tool for applying the snap buttons is available from the same source for less than $5.00, a

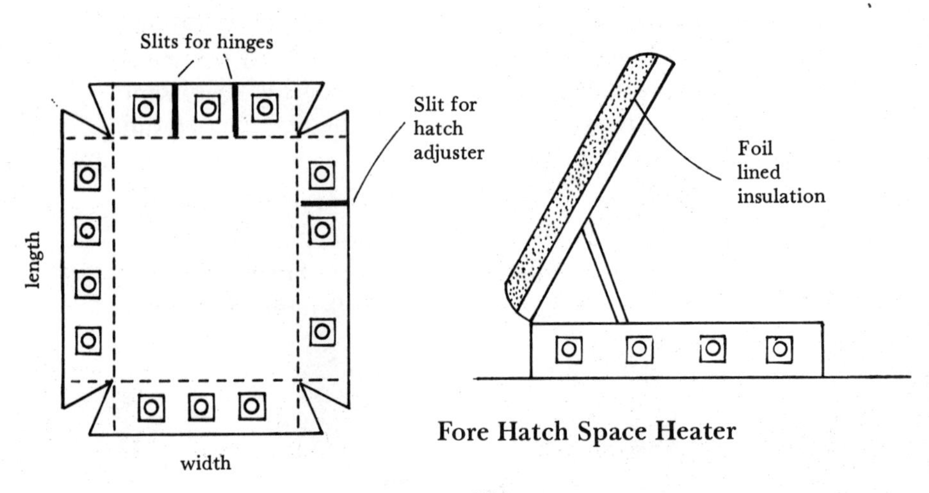

Fore Hatch Space Heater

worthy investment for a tool that will be used again and again on the solar boat.)

To increase the efficiency of this heater, glue a piece of foil-lined insulation to the hatch cover, shiny side out. This will reflect more of the sun's rays through the plastic cover and into the boat. In addition, when the hatch cover is closed, it will help hold the heat inside the boat.

List of Materials – Space Heater 1

1/2 yard .016 gauge transparent flexible vinyl film
8-16 marine canvas snap buttons and self-tapping studs
dacron thread
sail or upholstery needle

SPACE HEATER TO BE USED IN PORTHOLES

Use one or more of these basic space-heating devices to capture heat as the sun streams through the ports and to recirculate the air from inside the boat by natural convection. The air flowing from the top of the heater will be between 110 and 120 degrees F on sunny days.

To use the heater, open the port when sunlight begins to strike the collector box. At sundown, on cloudy days, and on the shady side of the boat, shut the port and cover it with an insulating cloth to keep the captured heat inside the boat.

The heater itself is simply a glass-covered insulated box that fits air-tight against the cabin sides. There is a blackened metal collector plate in the center of the box. It collects heat and separates the cool air from the hot air as it flows through the heater. The dimensions given are for a porthole that measures 8 × 15 inches and that is 6 inches from a 21-inch wide deck. It may be necessary to adjust these dimensions to suit your particular boat.

1. Cut all the pieces of plywood and insulation to the sizes shown in the list of materials.

2. Assemble the plywood box with waterproof glue and brads; then glue insulation to the bottom, end, and lower sides of the box.

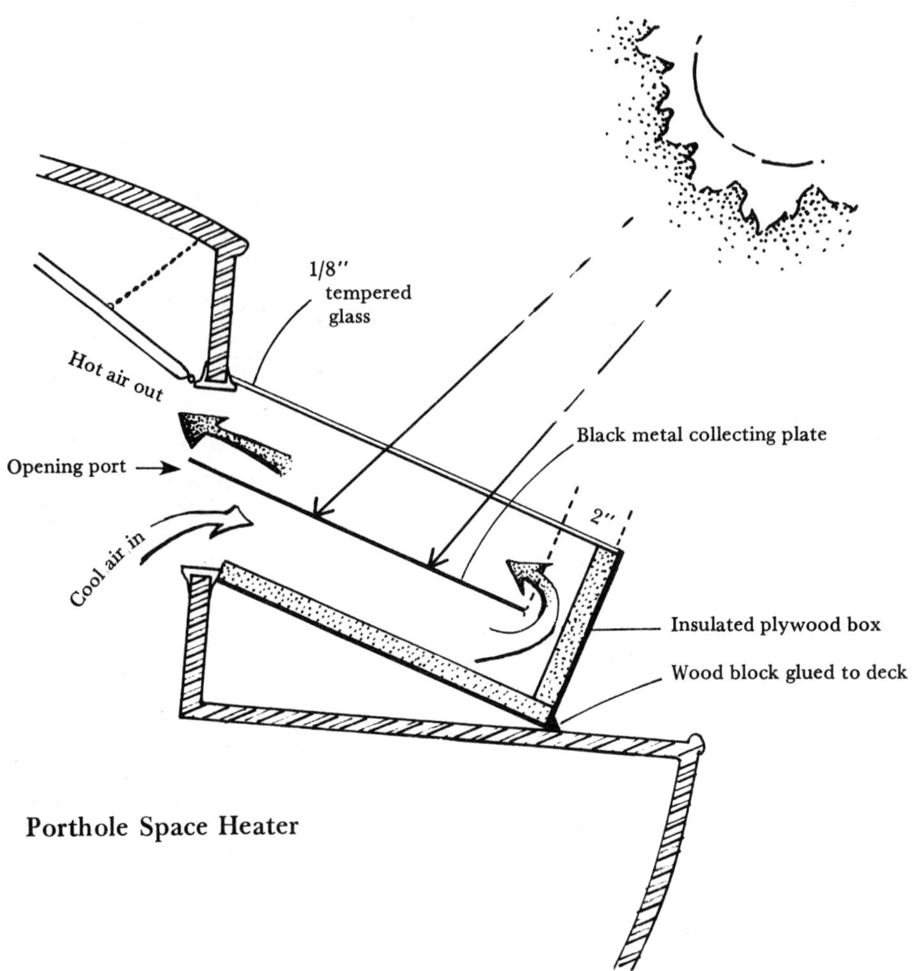

Porthole Space Heater

3. Cover the insulation with aluminum foil held in place with white glue, and paint the entire inside of the box and both sides of the metal collector plate with two coats of flat black stove enamel.

4. Put the collector plate in place and drill 4 holes through each side of the plate through the insulation and the plywood box. Bolt the plate in place through these holes.

5. Glue the remaining (top) side pieces in place, cover with aluminum foil, and paint flat black.

6. Set the tempered glass or plexiglass panel into a bed of waterproof caulking applied to the side and end pieces of insulation. When the caulking has cured, cover edges of glass with duct tape or flashing.

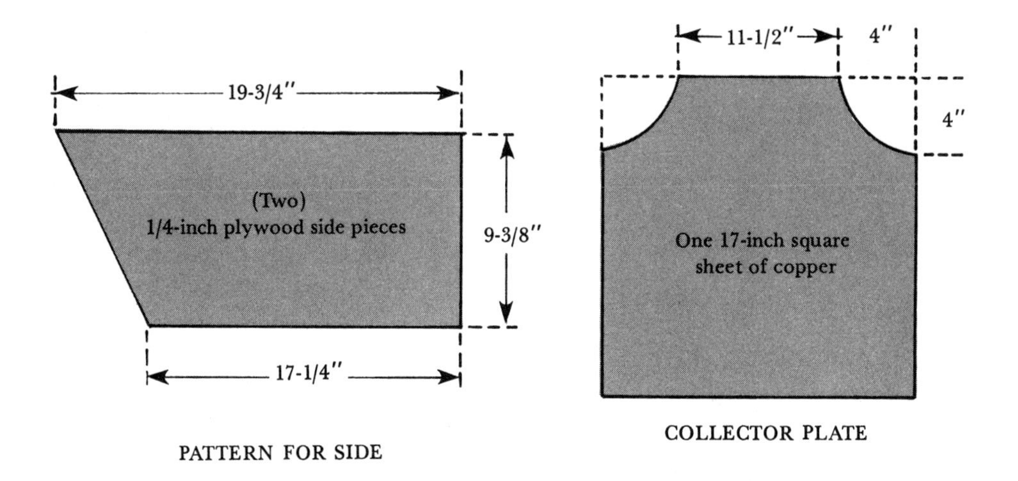

Detail, Porthole Space Heater

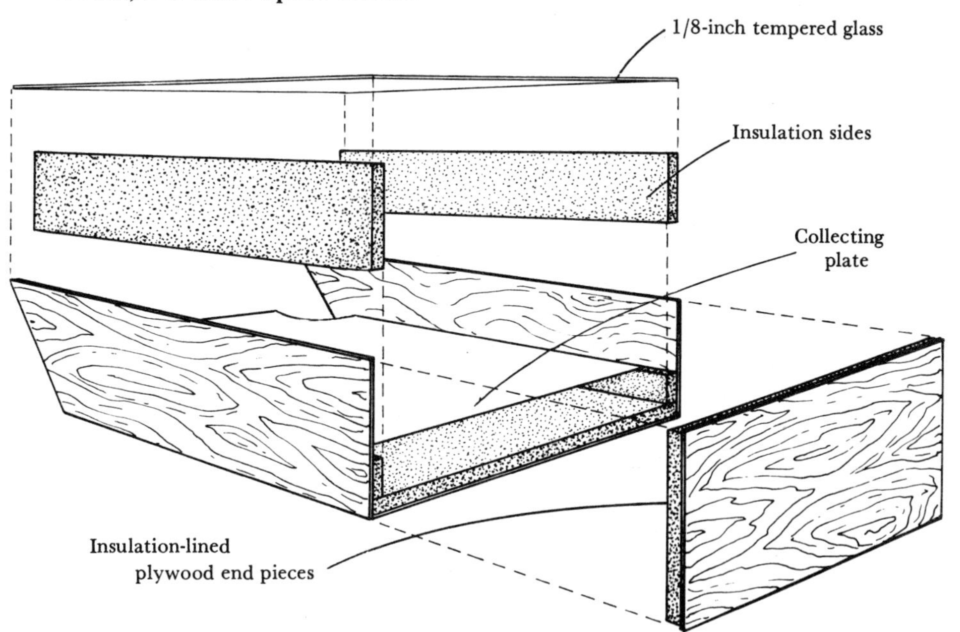

7. Reinforce the corners of the box with flashing or L-shaped aluminum, sand the box lightly and give it two or three coats of exterior enamel.

8. Cut a wooden wedge to fit the lower end of the heater and glue it to the deck.

The part of the heater that fits around the porthole will be held in place with duct tape so that it can be removed easily at the end of the winter season. If a more permanent installation is desired, it can be held in place with flashing.

The installation should be *air tight.*

List of Materials – Space Heater 2

waterproof caulking
waterproof glue
white glue
3/4-inch brads
10 ft. flashing
duct tape
aluminum foil
flat black stove enamel
1/8-inch tempered glass or plexiglass,
  cut 17 X 19-1/4 inches
sheet copper 17 X 17 inches,
  trimmed as shown in sketch

1/4-inch exterior grade or marine plywood, cut as follows:
1 end piece 9-3/8 X 17-1/2 inches
1 bottom piece 17-1/4 X 17 inches
2 side pieces as shown in sketch

1-inch rigid insulation board, cut as follows:
1 piece 9 X 17 inches (end)
2 pieces 16 X 2 inches (lower sides)
2 pieces 18 X 6 inches (upper sides)
1 piece 16 X 17 inches (bottom)

Solar Cookers

All solar cookers work best with a blackened iron pot with a pyrex lid. A 2-quart kettle is a good all-purpose size. It can be used for boiling, baking, steaming or frying. It will measure approximately 7 inches in diameter by 4 inches deep. A standard steamer basket will fit inside a kettle this size, the deep sides will contain splattered oil when frying, and it is large enough to prepare a one-dish meal for four adults. All three solar cookers and the insulated food keeper shown in this chapter are designed to be used with a kettle this size.

First, make certain key alterations in your cook pot.

1. Drill two holes 1/2 inch from the rim of the kettle and equidistant from each other.

2. Turn a nut with a lock washer onto a 1/4-inch bolt 4 inches long, push the bolt through the hole in the kettle and turn another nut with lock washer onto the end of the bolt.

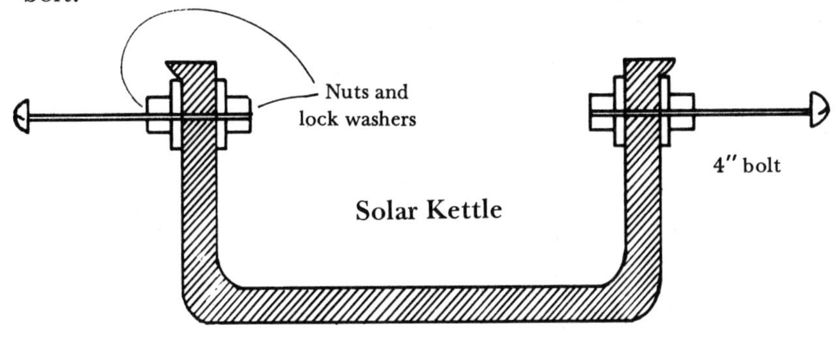

3. Repeat for the other side and tighten the nuts hard against the sides of the kettle.

The next step is to build a sturdy, adjustable stand to hold your solar kettle.

Our solar stand resembles a pair of boom crutches.

1. Cut 4 pieces of 1 X 2's two feet long and drill a hole 2 inches from one end of all four pieces.

2. Round all four corners of all the pieces and bolt them together in pairs.

3. A brace across the bottom of each crutch is made from heavy strap aluminum 16 inches long with holes drilled along the center line about 1 inch apart.

4. One end of the brace is fastened permanently to the crutch with a nut and bolt.

5. The other end is adjustable by means of a bolt and wing nut.

6. Another heavy aluminum brace 11 inches long is attached to each crutch to hold the two crutches the proper distance apart.

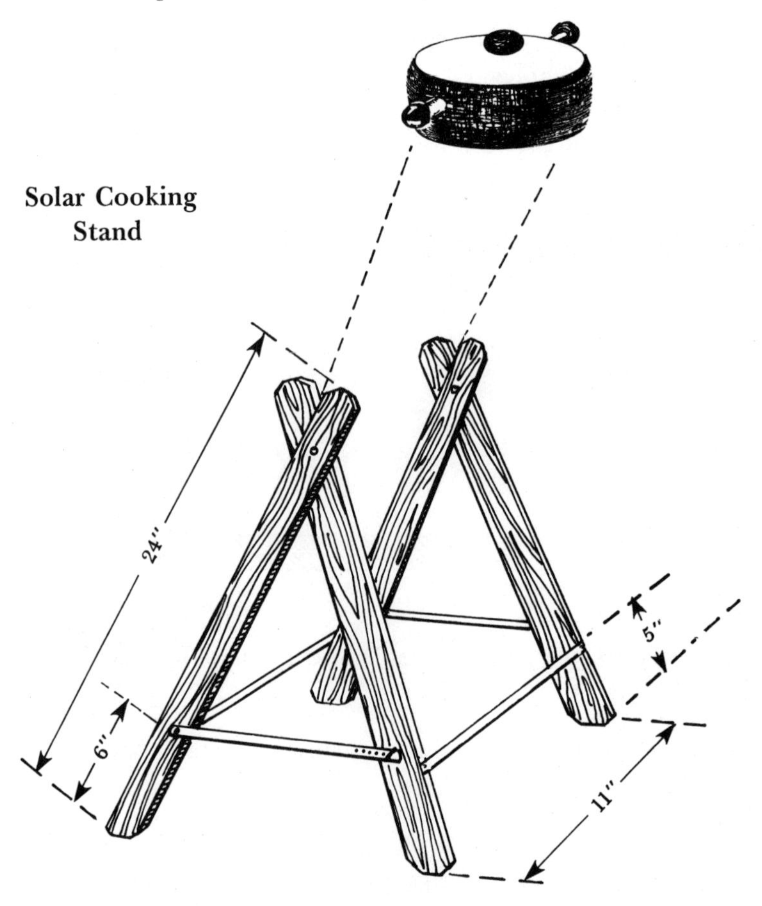

Solar Cooking Stand

7. One end of each brace is permanently attached to one leg of each crutch with a 1-inch wooden screw. The other end of each brace is attached to the crutch leg opposite it with a bolt and wing nut.

The entire assembly will fold flat for storage.

To use the stand, open each crutch to the proper height and fasten the crutch brace with a wing nut. Join the two crutches by opening the cross braces and fastening them in place with their wing nuts. Set the stand upright, and place your solar kettle in place with its protruding bolts resting in the "Y" formed by the top of the stand. Then, set the Fresnel lens cooker or the Folding Camp Cooker so that the sun is focused directly on the solar kettle.

List of Materials for Solar Kettle Stand

8 ft. 1 X 2 inch wood
4-1/2 ft. heavyweight flat aluminum strapping
2 (1-inch) wood screws
4 (1/4-inch) bolts 2 inches long
4 (1/4-inch) bolts 1-1/2 inches long
4 nuts and flat washers
4 wing nuts and flat washers

FRESNEL LENS COOKER

This cooker is a variation of the solar furnace. It will reach very high temperatures very quickly, so must be handled with great care. It is best used with a double boiler or steamer, or can be used for quick stir frying, provided protective gloves and dark glasses are worn constantly while cooking. This particular cooker folds flat for storage.

Begin by mounting a 12-inch square plastic Fresnel lens in a frame of 1 X 2's.

1. Cut a 1/4-inch deep groove in the center of the wooden frame to accept the lens.

2. Drill a hole through the center of both side pieces of the frame.

3. Insert a 1/4-inch bolt 2 inches long through these holes.

4. Assemble the frame with the lens in place, using waterproof glue and 1-inch brads.

The framed lens must now be mounted in a U-shaped frame. This frame can be bolted or clamped to a part of your standing rigging, or it can be attached to a plywood base which will allow you more freedom of movement. The U-shaped frame is also assembled from 1 X 2's, glued and screwed together. The side pieces are slotted 1 inch deep to accept the 1/4-inch bolts that protrude from the framed lens.

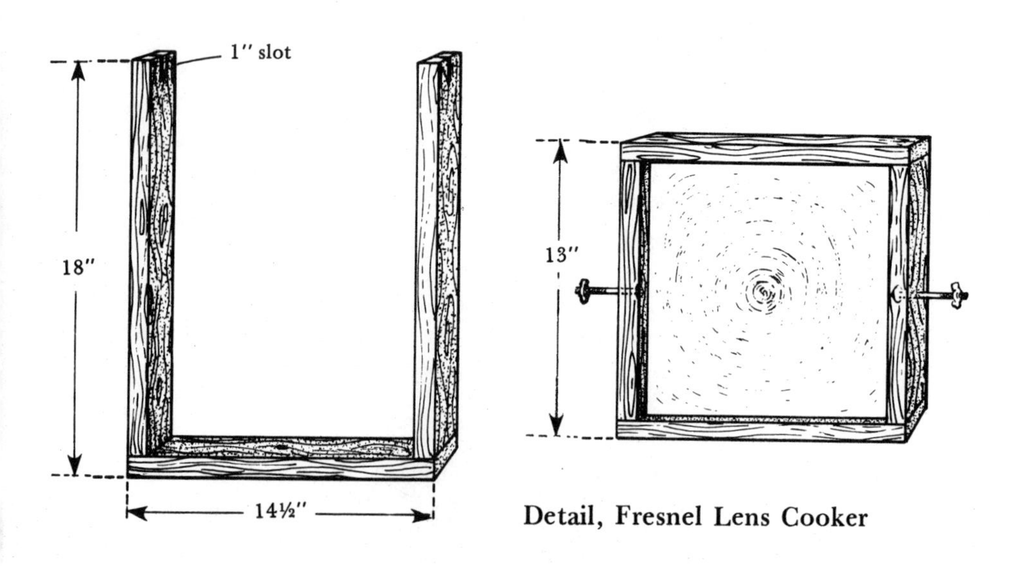

Detail, Fresnel Lens Cooker

Drill a hole in the center of the bottom part of the U-frame. Cut a base from 3/4-inch plywood 13 inches square or larger. Drill a hole in the center of the plywood base for a 1/4-inch bolt 2 inches long, and assemble the unit as shown.

Place your solar kettle in its stand before focusing the Fresnel lens cooker. The kettle should be about 15 inches from the focal point of the cooker. WEARING DARK GLASSES AND GLOVES, adjust the framed lens until the sun is focused on the front of the solar kettle. Tighten the wing nuts and stand back. Liquids will boil in less than 5 minutes. Use great care when focusing the lens. Wood will char and burn and paper will burst instantly into flame if placed at the focal point of the lens.

List of Materials
for Fresnel Lens Cooker

1 Fresnel lens 12 inches square (available at local hobby shops or from Edmund Scientific)
9 ft. 1 X 2 inch wood
3/4-inch plywood 13 X 13 inches or larger
3 (1/4-inch) bolts 2 inches long
3 flat washers and wing nuts
4 (1-1/2-inch) wood screws
1-inch brads
waterproof glue
exterior paint or varnish

Fresnel Lens Cooker

THE SOLAR OVEN

This solar oven is very easy to construct and to use. The basis for this oven is a recycled 5-gallon round metal can. Used cans such as these usually can be obtained free from a building contractor.

1. Begin by cleaning the can thoroughly. Remember that it will reach high temperatures, so traces of oil or chemicals can be dangerous if they are not removed.
2. Cut a 7-inch circle in the center of the lid and snip the edge of the remaining circle every inch, cutting 1/4 inches deep.
3. Cut an 11-inch circle from the foil-backed spun glass insulation and glue it to the bottom of the can, foil side up.
4. Cut another piece of insulation 11 X 35-1/2 inches and glue it to the sides of the can, foil side out.

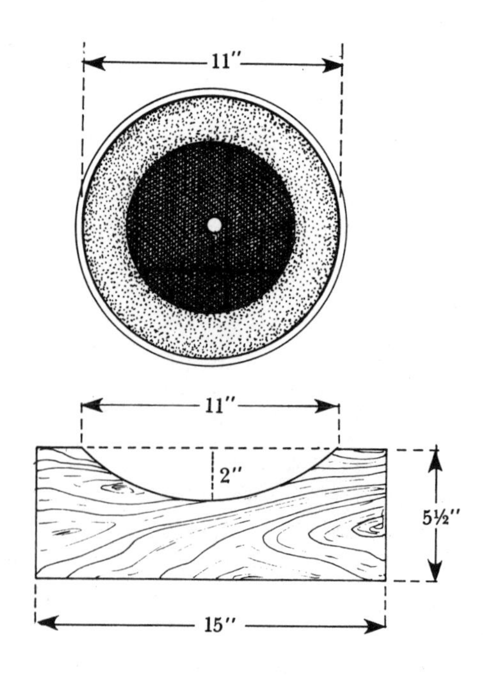

Solar Oven

5. Cut a rectangle of heavy gauge aluminum 7 X 10 inches.

6. Slip the rectangle into the can and put the lid in place, clamping it securely.

7. With pliers, turn the snipped edges of the cut circle *in* over the insulation.

8. Sand the outside of the can lightly and give it two coats of flat black stove enamel.

9. Have a piece of double strength oven glass cut in a circle 11-1/2 inches in diameter with a small hole drilled in the center. Attach a drawer-pull. Drive a small nail in the center of the drawer-pull as an aid in focusing.

10. Cut four squares of aluminum 11 X 11 inches and drill a 1/4-inch hole in two corners of each square. In one of the squares, drill a third hole in the center between these two holes. These squares will be mounted around the open end of the can as shown in the sketch.

11. Attach four hinges equidistant around the can with self-tapping screws. Attach the hinges to the aluminum squares with nuts and bolts. They should be positioned so that they will fold flat across the open end of the can.

12. Turn a self-tapping No. 8 screw into the top center back of the can. Leave two or three threads exposed.

13. The final step is to cut a 2 X 6 inch piece of wood 15 inches long. Cut a hollow into the center of one side of the wood 2 inches deep and 11 inches wide.

To use the oven, set the can into the block of wood so that it faces the sun. Open the top reflector and run a piece of galvanized wire from its center to the screw at the top back of the can. Open the side reflectors and run a piece of wire between the top reflector and each side reflector. Last, open the bottom reflector, also attaching it to the side reflectors with wire. The oven is properly positioned when the sunlight is reflected on the center of the blackened insulation inside the can. Experiment until it is just right. Lay an oven thermometer on the aluminum shelf, put the glass cover in place and preheat your oven.

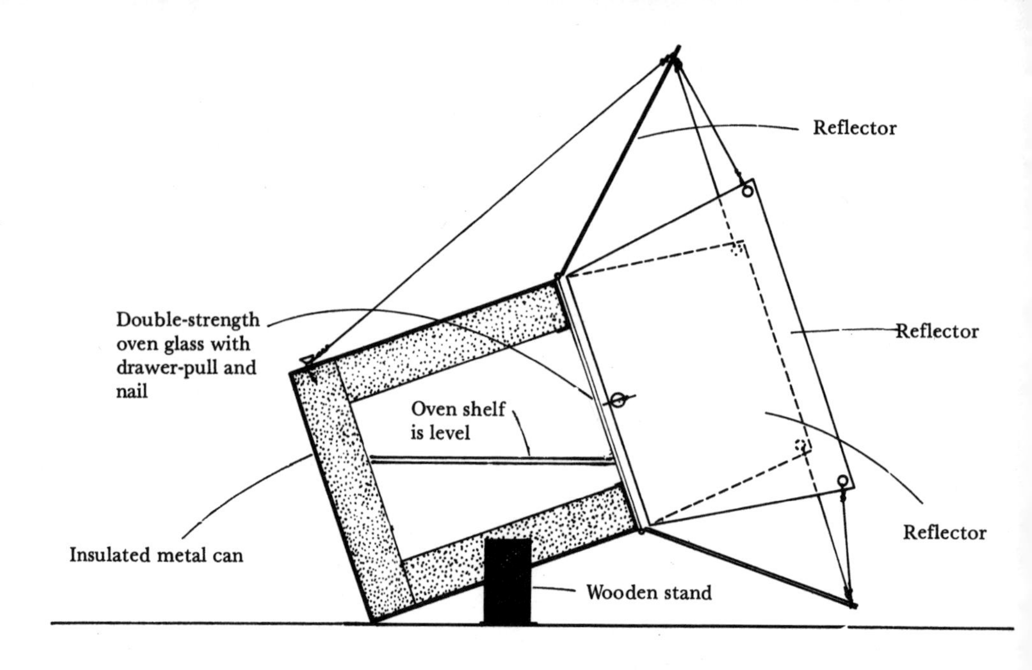

This oven should reach 350 degrees F in 15 to 20 minutes, then the temperature will continue to climb very slowly. To lower the oven temperature, prop the top of the glass cover open slightly.

List of Materials for Solar Oven

5-gallon round metal can with cover
4 sq. ft. foil-backed spun glass insulation
4 pieces sheet aluminum cut 11 X 11 inches each
7 X 10 inch heavy gauge sheet aluminum
4 hinges
8 self-tapping screws
8 nuts and bolts
1 No. 8 pan head self-tapping screw 1-1/2 inches long
1 drawer-pull
Pliobond cement
11-1/2 inch circle double strength oven glass
35-inch sealing strip for glass
12 ft. galvanized wire
15 X 2 X 6 inch wood
flat black stove enamel
oven thermometer

FOLDING CAMP COOKER

This cooker is a folding parabolic collector, constructed from cardboard covered with aluminum foil or aluminized mylar film. If desired, it could be made from sheet aluminum.

1. Begin by drawing a pattern on a piece of butcher paper or newspaper, using the dimensions shown in the sketch as a guide. You will need 9 sq. ft. of heavy cardboard and 9 sq. ft. of aluminum foil.

2. Glue the foil, shiny side up, to the cardboard with white glue. Keep it as wrinkle-free as possible.

3. When the glue is dry, trace your pattern directly onto the foil as shown in the sketch, and cut out the pattern carefully, using a single edge razor blade or exacto knife.

4. Punch the hole in the point of each triangle with a paper punch. You will need 16 of these triangles. Stack them together once they are cut out.

The next step is to construct a stand.

5. Cut a 4-inch square from a piece of 1 X 6 inch wood. In the center of one side, cut a notch 1-inch wide by 1-1/2 inches deep, as shown.

6. Cut one leg 1 X 2 X 21 inches and round all the edges.

7. Insert the leg in the notch and drill a hole through the block and the leg to accommodate a 1/4-inch bolt 4-1/2 inches long.

8. Insert the bolt and be sure the leg swivels freely. Then, turn on the wing nut over a flat washer.

9. The opposite leg is made from 1 X 2's glued and nailed together as shown. Round the edges, put in position and drill a hole through both sides of the leg and the square block, large enough to accommodate a 1/4-inch bolt 6 inches long.

10. Insert the bolt, washer and wing nut.

11. Drill a 1/4-inch hole in the center of the wooden block and insert a bolt 2 inches long.

12. Drop the stack of previously cut triangles on the center bolt, put a flat washer on top and turn a wing nut onto the bolt. Do not tighten it down. Leave room for the triangles to swivel.

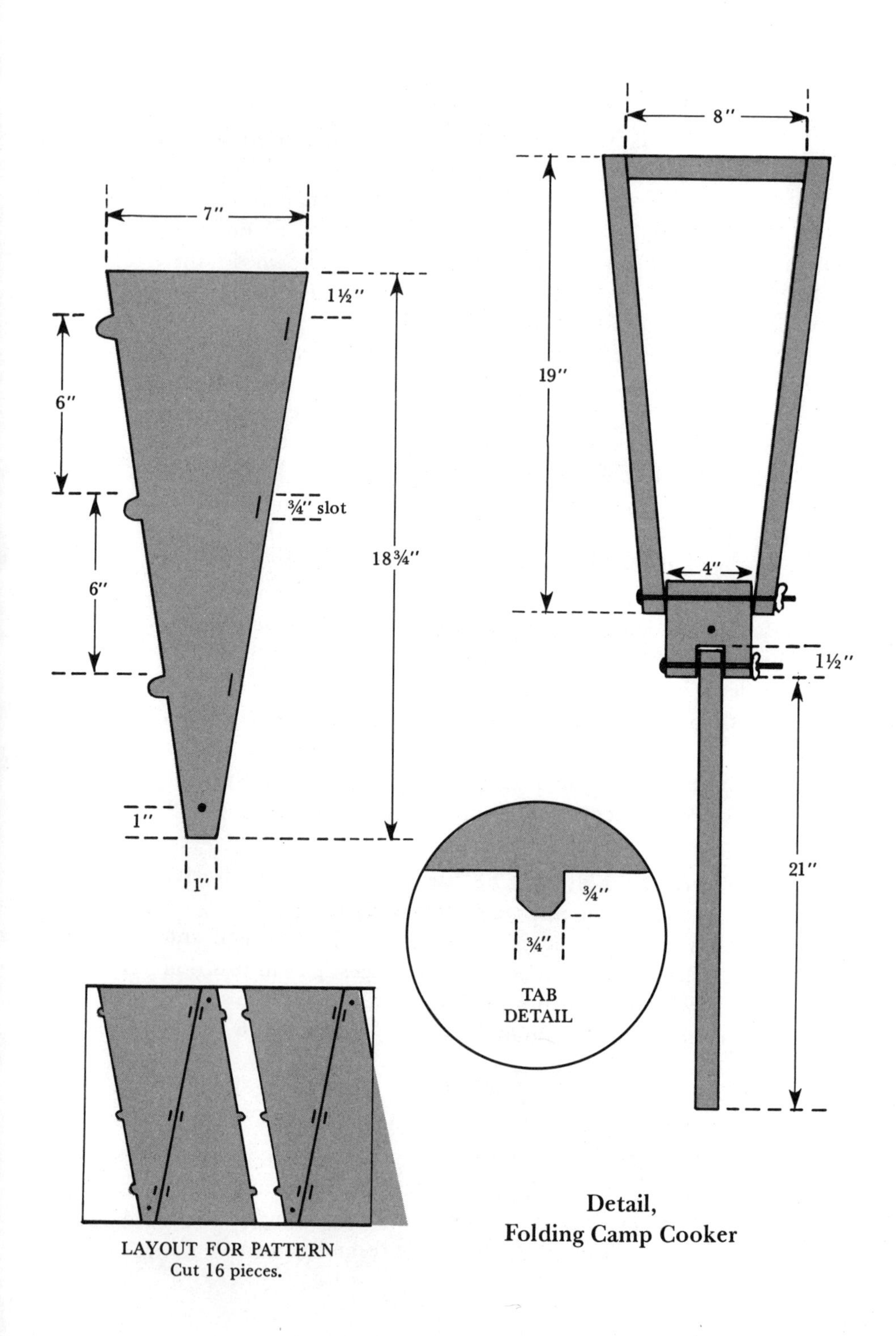

**Detail,
Folding Camp Cooker**

13. To use this cooker, lay it with the legs on the bottom and the triangles on the top. Beginning with the bottom triangle, swivel it to the left until you can see the slots on one side. Push the tabs on the next triangle *down* into these slots. Repeat this process until you get to the top triangle. Then push the tabs on the bottom triangle *up* into the slots of the top triangle. Tighten the center wing nut.

14. Adjust the legs with the double leg in front and the single leg at the back. When the reflector is aimed so that the sun's rays intersect at a hot spot 10 to 11 inches out from its center, it is in the proper position.

Next, put your solar kettle into this hot spot. You may have to modify your cooking stand somewhat by using longer braces, or you might position the reflector so that the solar kettle can hang from a halyard. This solar stove will not reach the extremely high temperatures you can expect from the Fresnel lens cooker, but it will boil water in about 15 minutes or fry bacon crisp in 10 minutes on a sunny day.

Remember when using any solar cooker that wind, clouds, or dirty reflective surfaces are the enemy. Keep all reflective surfaces bright and shiny and protect your cooker from the wind, when necessary. You can't do much about the sunny day that suddenly clouds up and stays that way, except to retreat to your alternate cooking methods. More tips on solar cookery are contained in Chapter 9.

List of Materials for Folding Camp Cooker

9 sq. ft. heavyweight cardboard
9 sq. ft. aluminum foil or aluminized mylar
1 X 6 X 4 inch wood
6 ft. 1 X 2 inch wood
1-inch brads
white glue
3 (1/4-inch) bolts 2 inches long
3 (1/4-inch) bolts 4-1/2 inches long
3 (1/4-inch) bolts 6 inches long
3 flat washers/wing nuts

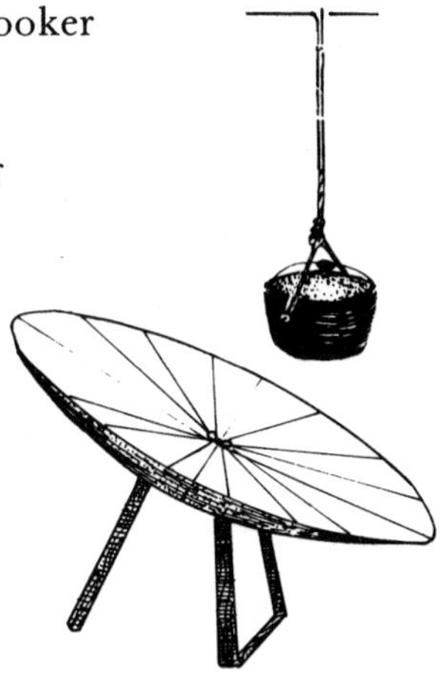

INSULATED FOOD KEEPER

To use this insulated box, food must first be brought to the boiling point, 212 degrees F or higher, then quickly placed into the box where its own heat is retained to finish cooking. Use a good quality cooking thermometer and follow these instructions carefully to avoid culturing bacteria or other harmful toxins.

This box is built to accommodate the solar kettle we've already discussed. Adjust the dimensions to suit your largest or most used kettle.

1. Glue aluminum foil, shiny side out, to one side of the insulation. Then cut out all the pieces of plywood and foil-backed insulation to the sizes shown in the List of Materials.
2. Construct a box from the plywood as shown in the sketch, using wood glue and 3/4-inch brads.
3. Glue the pieces of insulation inside the plywood box, first the sides, then the front, and finally the bottom.
4. Leaving 2-1/4-inch clearance on all sides, glue the top insulation to the top plywood.
5. Drive a few brads into the top insulation from the plywood top.
6. Finally, frame the edges of the lid with L-shaped wood moulding, glued in place with epoxy glue.

When using this box, preheat the solar kettle to at least 212 degrees F and put it into the box. Stuff an old towel or a wadded-up newspaper into the box, filling the spaces around the kettle, and push the lid tightly into place. Recipes and more information on using this insulated box are included in Chapter 9.

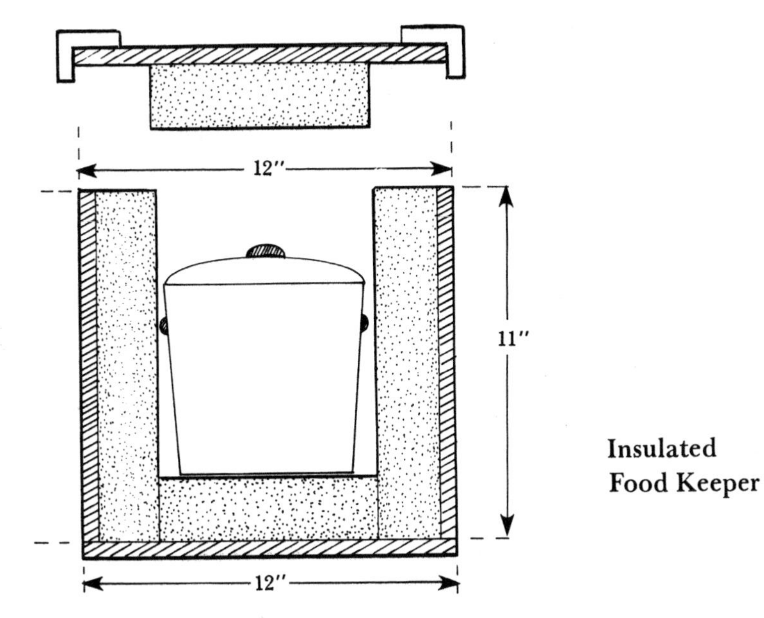

Insulated
Food Keeper

List of Materials
for Insulated Food Keeper

1/4-inch plywood, cut as follows:
- 2 pieces 12-1/2 X 12-1/2 inches (top and bottom)
- 2 pieces 11 X 12 inches (sides)
- 2 pieces 11 X 12-1/2 inches (front and back)

2-inch thick rigid foam insulation, cut as follows:
- 2 pieces 8 X 8 inches (top and bottom)
- 2 pieces 11 X 12 inches (sides)
- 2 pieces 11 X 8 inches (front and back)

4 ft. L-shaped wood moulding
white glue
3/4-inch brads
epoxy glue
thermometer

12

Putting It All Together Energy Odds and Ends

Now that your head is swimming with all the things that go into solarizing your boat, where do you begin?

Keep in mind that the Solar Boat is the perfect cruising boat. It is equipped for offshore cruising, where food, water, fuel, spare parts and shore power receptacles may be in short supply or nonexistent. The Solar Boat uses solar energy wherever it is suited to the job at hand, with other natural energy sources as backups to the sun. The Solar Boat also uses conventional energy sources when they are more appropriate for the job, or as an additional backup, but in the most frugal manner possible.

Everything you do to bring your boat up to this level of independence will not only increase the value of your boat, but will repay you for your efforts in *cash* savings in fuel, utilities and marina fees.

Begin by thoroughly evaluating your electrical systems, both AC and DC. Compare the amount of electricity you use with the amount of electricity you are able to generate and to store. Review and reduce those electrical demands in every way you can without sacrificing your comfort or life style. Satisfy your electrical needs, but don't make unnecessary demands on your electrical system.

Label each piece of electrical equipment on board with the amount of amperage it uses.

Install as many solar generators or micro-generators as you can.

Add other free fuel generators (wind or water) to supplement your solar arrays.

Set up a thorough, regular maintenance schedule for conventional alternators/generators as well as free fuel generators and all electrical equipment, in order to keep them in peak operating condition.

Our greatest energy demand is for heat, and the easiest way to harness the sun is to use it to generate heat. Begin by building a solar water heater for your boat. Keep it as simple as possible in relation to your personal needs. Try one of the more primitive solar water heaters first. This will quickly convince you not only of the great power of the sun, but also the ease with which its energy can be used to generate heat. In cold climates, be sure to provide auxiliary heating and freeze protection for your solar water heater.

Next, using the design principles outlined earlier in this book, turn your boat into a combination solar heat collector/trap/storehouse.

Be sure the color of your boat complements the climate in which it will be used.

Whenever possible, the largest glass areas of your boat should point within 15 degrees of due south.

Seal your boat tightly. Caulk, rebed and weatherstrip wherever necessary. Be sure outside cabin doors, hatch covers, windows and ports, windscoops, vents and chimneys have air- and water-tight seals.

Insulate as much as you possibly can: ceilings, cabin sides, lockers, under decks and cabin soles, hot water storage tanks, water pipes, etc.

Add double glass to windows, ports and glass sliding doors to prevent condensation and heat loss through the glass.

Use reflective film on glass areas when appropriate.

Use insulating cloths on windows, ports, and hatches, and cover them with insulating drapes.

Use exterior shutters, when possible.

Add thermal mass wherever you can.

Protect the main entry as much as possible. Use an awning with attached side curtains, a dodger, or best of all, a complete clear vinyl cockpit enclosure. This will form a passive solar heating system through the "greenhouse effect."

In colder climates, add auxiliary passive solar heating devices, such as the solar window or hatch heater. In extremely cold weather, when further backup heating is needed, use a heater that burns a renewable-type fuel.

For warm climates, be sure that all windows and ports open for ventilation, and that openings are available on all four sides of the boat.

In warm weather, provide a large harbor awning or series of awnings, which will shade the entire boat.

In extremely hot weather, choose an offshore anchorage in preference to a berth dockside.

When necessary, add auxiliary cooling devices, such as wind chutes, water trickling over the cabin top, small fans, or a larger hatch-mounted fan.

An air conditioner should be your last resort, but if you should choose to use one, select the air conditioner with the highest EER rating.

Live as close to the sun as you can in all areas of your life. Sun dry foods when they are in season and plentiful, then store them for future use.

Use your solar cooker and insulated box as often as you find convenient. You can make solar tea every day that the sun shines.

Follow the solar recipes in this book and experiment with some of your own.

But, most importantly, THINK SOLAR. Read about solar energy, write your congressman, support solar programs, tell your friends and encourage your children to learn as much as possible about the sun.

Keep yourself posted on the latest solar developments. Don't be afraid to try something new. Now that you are armed with a broad, general knowledge of solar technology, it will be difficult for you to be fooled by a solar con artist.

A small investment in solar equipment now is a big investment in the future – yours, your children's and all the generations to come.

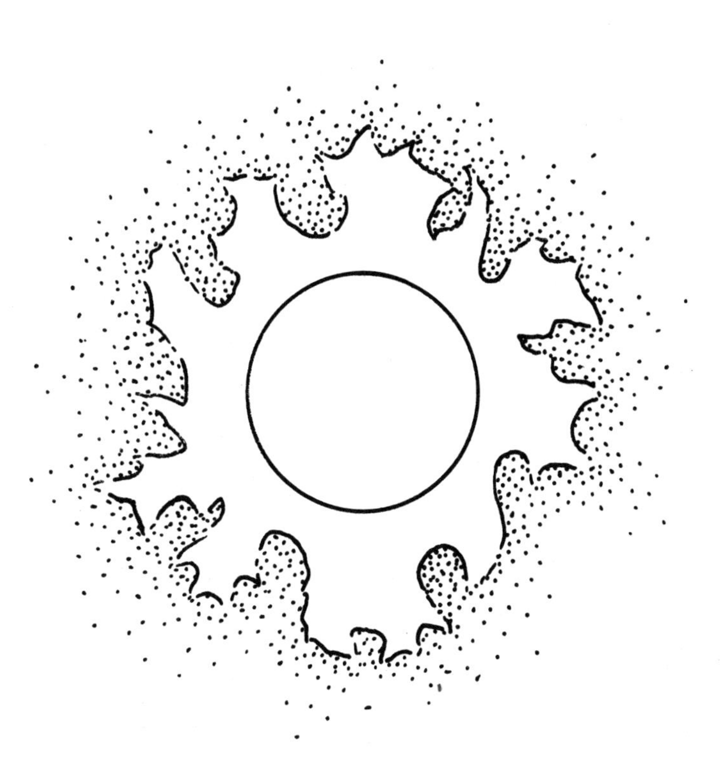

References

CHAPTER 1 / THE BASICS – THOSE FORGETTABLE FACTS

1. Floyd Hickok, *Handbook of Solar and Wind Energy,* A Cahners Special Report, CBI Publishing Co., Inc., Boston, Massachusetts, 1975, p. 12.

2. Farrington Daniels, *Direct Use of the Sun's Energy,* Ballantine Books, New York, New York, 1974, pp. 37-42.

3. *Fun with Fresnel Lenses,* No. 9053, Edmund Scientific Co., Barrinton, New Jersey, 1971, p. 5.

4. Daniels, pp. 208-215.

FURTHER SOURCES

Morton Mott-Smith, *The Concepts of Energy Simply Explained,* Dover Publications, Inc., New York, New York, 1964.

Mother Earth News, Handbook of Homemade Power, Bantam Books, Inc., New York, New York, 1974.

D.S. Halacy, Jr., *Fun With the Sun,* The Macmillan Company, New York, New York, 1959.

CHAPTER 2 / EVALUATING YOUR ELECTRICAL NEEDS

1. Amory B. Lovins, "Involvement Day Keynote Speeches – Seattle," Special Report, Supplement to Calypso Log, Volume 5, No. 1, March 1978, p. 2.

2. Conrad Miller, *Your Boat's Electrical System, Motor Boat and Sailing Book,* New York, New York, 1973, p. 20.

3. Jan and Bill Moeller, *Living Aboard,* International Marine Publishing Co., Camden, Maine, 1977, p. 20.

4. *Mother Earth News, Handbook of Homemade Power,* Bantam Books, Inc., New York, New York, 1974, p. 180.

5. Miller, pp. 37-40.

CHAPTER 3 / FREE FUEL GENERATORS: SUN – WIND – WATER

The following sources are very useful on using the sun, wind, and water as fuel:

Farrington Daniels, *Direct Use of the Sun's Energy,* Ballantine Books, New York, New York, 1975.

Floyd Hickok, *Handbook of Solar and Wind Energy,* A Cahners Special Report, CBI Publishing Co., Inc., Boston, Massachusetts, 1975.

Sensor Technology, Solar Electric Systems and Cells, Sensor Technology, Inc., Chatsworth, California.

Solar Energy Modules, Sensor Technology for Recreational and Survival Equipment, Sensor Technology, Inc., Chatsworth, California.

Keith Taylor, "Tabarly Wins Solo Race Again," *Sail,* August 1976.

Solar Electricity, Solarex Corp., Rockville, Maryland.

Morton Mott-Smith, *The Concept of Energy Simply Explained,* Dover Publications, Inc., New York, New York, 1964.

Spilling Consult AG, enclosures, SW 152e/SW 180e, MA 355-2, Information sheet 45-76.

Andy Ross, *Stirling Cycle Engines, Solar Engines,* Phoenix, Arizona, 1977.

The Radio Engineering Handbook, Keith Henney, Editor, Krieger Publishing Co., Melbourne, Florida, 1959.

Vega-C Research data sheet, "See Power With Sea Power," Vega-C Research, P.O. Box 568, Miami, Florida 33133.

CHAPTER 4 / REDUCING YOUR ELECTRICAL DEMANDS

1. Jan and Bill Moeller, *Living Aboard,* International Marine Publishing Co., Camden, Maine 04843, 1977, p. 98.

CHAPTER 5 / MARINE REFRIGERATION: SOLID STATE TO ICY BALL

1. Janet Groene, *The Galley Book,* David McKay Co., Inc., New York, New York, 1977, p. 51.

2. Jan and Bill Moeller, *Living Aboard,* International Marine Publishing Co., Camden, Maine 04843, 1977, p. 125.

3. Groene, p. 64.

4. Don McCoun, *Ice-Less Ice,* McCoun Enterprises, Rockville, Nebraska, 1978, p. 1.

5. Instructions for Crosley Icy Ball Refrigerator, furnished by Elizabeth Brown, *Air Conditioning, Heating and Refrigeration News,* P.O. Box 2600, Troy, Michigan 48007, Sept. 1977, and E.W. Bottum, President, Solar Research Division of Refrigeration Research, Inc., 525 N. 5th St., Brighton, Michigan 48116, Sept. 1977.

6. Farrington Daniels, *Direct Use of the Sun's Energy,* Ballantine Books, Inc., New York, New York, 1974, pp. 167-170.

CHAPTER 6 / SOLAR HOT WATER: HOW TO GET IT – HOW TO KEEP IT

1. Farrington Daniels, *Direct Use of the Sun's Energy,* Ballantine Books, Inc., New York, New York, 1974, p. 75.

2. *Miami Herald,* Sunday, July 30, 1978, "Warming Trend," p. 19H.

3. Bruce Anderson with Michael Riordan, *The Solar Home Book,* Brick House Publishing Co., Andover, Massachusetts, 1976, p. 210.

4. Bruce Anderson with Michael Riordan, pp. 216-217.

5. *Ibid,* pp. 223-225.

6. *Solar Utilization News, Solar Hot Water Systems,* November 1977, The Alternate Energy Institute, Estes Park, Colorado, p. 17.

7. "The Khanh Solar Water Heater," *Mother Earth News,* No. 45, Hendersonville, North Carolina, May/June 1977, p. 124. Dike Mason, "A Simple Solar-Heated Shower," *Mother Earth News,* No. 46, July/August 1977, p. 64. Miles K. Free, "Recycle a Refrigerator into a Solar Water Heater," *Mother Earth News,* No. 48, November/December 1977, pp. 108-109. Doyle Akers, "Doyle Akers $30 Homestead Solar Water Heater," *Mother Earth News,* No. 51, May/June 1978, pp. 122-124.

CHAPTER 7 / PUT THE HEAT OF THE SUN TO WORK ON YOUR BOAT

1. Bruce Anderson with Michael Riordan, *The Solar Home Book,* Brick House Publishing Co., Andover, Massachusetts, 1976, pp. 77-81.

2. Farrington Daniels, *Direct Use of the Sun's Energy,* Ballantine Books, New York, New York, 1975, pp. 57-61.

3. Bruce Anderson with Michael Riordan, p. 86.

4. Boughton Cobb, Jr., *Fiberglass Boats, Construction and Maintenance,* Yachting Publishing Corp., 1965, p. 103.

CHAPTER 8 / THE NITTY GRITTY OF DIRECT SOLAR HEATING AND COOLING

1. Jan and Bill Moeller, *Living Aboard,* International Marine Publishing Co., Camden, Maine 04843, 1977, pp. 180-181.

2. 3M Company, P.O. Box 33800, St. Paul, Minnesota 55133, Design Guide Specification, "Scotchtint" Sun Control Film A-20, pp. 1 and 2.

3. *Solar Utilization News,* Solar Window Heater Plans, May 1978, The Alternate Energy Institute, Estes Park, Colorado, pp. 6, 16, 17.

4. Rick Heiman, *Solar Concepts,* Box 263, Stafford, Texas 77477, Solar Heater Options and Applications, pp. 1-4.

5. Floyd Hickok, *Handbook of Solar and Wind Energy,* A Cahners Special Report, CBI, Boston, Massachusetts, 1975, p. 2.

CHAPTER 9 / SOLAR FOOD AND DRINK FROM COOKERS TO THE SOLAR STILL

1. *Solar Cooker,* Davis Instruments Corp., 857 Thornton St., San Leandro, California 94577, 1978, pp. 1-4.

2. Mother's Solar Powered Hot Dog Cooker, *Mother Earth News,* No. 50, March/April 1978, p. 32.

3. Mother Tests the Solar Chef, *Mother Earth News,* No. 44, March/April 1977, p. 90h.

4. *Fun with Fresnel Lenses,* No. 9053, Edmund Scientific Co., 1971, pp. 5-7.

5. Now Here's New Kind of Stove, *Miami Herald,* Thursday, September 14, 1978.

6. Dr. C.G. Abbot, How to Build a Solar Cooker, *Mother Earth News,* No. 45, May/June 1977, reprinted from *Science and Invention,* June 1923.

7. Robert Roth, *All About Heat Pipes,* Edmund Scientific Co., Barrinton, New Jersey 08007, 1973, pp. 9-11.

8. *Build It Better Yourself,* Rodale Press, Emmaus, Pennsylvania, 1977, pp. 60-61.

9. Drying Foods at Home, U.S. Department of Agriculture, *Home and Garden* Bulletin No. 217, January 1977, pp. 8, 9.

10. *Home Preservation of Fishery Products,* reprinted from U.S. Department of Interior Pamphlet no longer in print, in Windvane, Volume 1, No. 2, October 1976, pp. 72-75.

11. Farrington Daniels, p. 6.

12. Julia F. Morton, *Wild Plants for Survival in South Florida,* Trend House, Tampa, Florida, 1974, p. 72.

CHAPTER 10 / REDUCING YOUR DOCKSIDE ENERGY DEMANDS

1. Conrad Miller, *Your Boat's Electrical System, Motor Boating and Sailing Books,* New York, New York 10019, 1973, pp. 65-87.

2. Jan and Bill Moeller, *Living Aboard,* International Marine Publishing Co., Camden, Maine 04843, 1977, p. 137-138.

Index

May we introduce
other Ten Speed Books
you will find useful . . .

Sailing the Farm
by Ken Neumeyer

Sailing the Farm sets out to equip the sailor with the information needed to turn a boat into a survival retreat. Topics covered include – Gathering and preparing edible seaweeds; Earning a living with your boat; Delicious no-fuel recipes; Survival tools; Long-term food storage; On-board money-making crafts; Vegetable gardening; Preparing for bad political and economic weather. Fully illustrated with diagrams and sketches, this unique book calls the sailor in each of us back to the sea . . . 6 x 9 inches, 256 pages, $7.95 paper, $14.95 cloth.

The Solar Cat Book
by Jim Augustyn

The Solar Cat Book is the first really humorous book to come out of the alternative energy movement. Jim Augustyn manages to blend humor with a remarkable amount of good, common-sense solar information, and comes up with what The Sierra Club Bulletin calls, ". . . a rare work: informative and funny." 6 x 9 inches, 96 pages, illustrated, $4.95 paper.

Before You Build
A Preconstruction Guide
by Robert Roskind, The Owner Builder Center

For most people, building or buying a house is the largest single expenditure made in a lifetime. *Before You Build* provides potential owner-builders with the essential tool of a comprehensive guidebook and housebuilding checklist, based on the experience of the recognized leaders in the field – The Owner Builder Center in Berkeley, California. 8½ x 11 inches, 192 pages, $7.95 paper.

The Manual for the home and farm production of Alcohol Fuel
by Stephen W. Mathewson

Here, in clear, straightforward language are chapters covering U.S. government regulations, fuel theory, modifying engines to burn pure alcohol, gasohol, water injection, economic considerations, malting, mashing, fermentation, distillation, and much, much more. A must for anyone interested in making fuel for home and farm use. 5½ x 8½ inches, 224 pages, $7.95 paper.

You will find them in your bookstore or library,
or you can send for our free catalog:

Ten Speed Press, Box 7123, Berkeley, CA 94707